Project Portfolio Management in Theory and Practice

Thirty Case Studies from around the World

Best Practices and Advances in Program Management Series

Series Editor
Ginger Levin

Project Portfolio Management in Theory and Practice

Thirty Case Studies from around the World

Jamal Moustafaev, MBA, PMP

CRC Press
Taylor & Francis Group
Boca Raton   London   New York

CRC Press is an imprint of the
Taylor & Francis Group, an **Informa** business

AN AUERBACH BOOK

CRC Press
Taylor & Francis Group
6000 Broken Sound Parkway NW, Suite 300
Boca Raton, FL 33487-2742

Printed on acid-free paper
Version Date: 20160419

International Standard Book Number-13: 978-1-4987-6924-2 (Hardback)

Library of Congress Cataloging-in-Publication Data

Names: Moustafaev, Jamal, 1973- author.
Title: Project portfolio management in theory and practice : thirty case studies from around the world / Jamal Moustafaev.
Description: Boca Raton, FL : CRC Press, 2017. | Includes bibliographical references and index.
Identifiers: LCCN 2016011453 | ISBN 9781498769242 (hardcover : alk. paper)
Subjects: LCSH: Project management--Case studies. | Project management--Finance. | Portfolio management.
Classification: LCC HD69.P75 M692 2017 | DDC 658.4/04--dc23
LC record available at https://lccn.loc.gov/2016011453

Visit the Taylor & Francis Web site at
http://www.taylorandfrancis.com

and the CRC Press Web site at
http://www.crcpress.com

To my three-year-old son Shamil, who, save for the regular comic relief, provided no assistance whatsoever in the writing of this book.

Contents

SECTION II THE APPLICATION: INDUSTRY CASE STUDIES

SECTION III   SUMMARIZING IT ALL

Preface: Why Another Book on Project Portfolio Management?

PPM is the science and the art of selecting the best projects for the organization and maintenance of the project pipeline subject to internal and external constraints.

An Awkward Conversation with a CEO

I remember a consulting engagement that happened several years ago that became an inspiration for writing this book. I was invited to a meeting with several high-ranking executives of a very large port authority. All I knew before the meeting was that they seemed to have some project-related issues they wanted to discuss with me.

We sat down in a posh conference room with the CEO, COO, and several vice-presidents and commenced our discussion about the value of eliciting detailed requirements, planning, monitoring, and control of their projects. I noticed that the CEO of the company, while really eager to participate, looked like he had something else, something very important on his mind. Finally, he found a moment of quiet in the room and the following conversation took place:

> CEO: There is another problem and I am not sure if it is within your domain of expertise...
>
> Me: I am listening!
>
> CEO: I constantly get complains from our middle management that they do not have enough resources to deliver all of their projects. The way I see it, I have several options:

- I ignore their requests and tell them to roll up the proverbial sleeves and work harder or
- I should either provide them with more resources—both human and financial—or I need to cut some of their projects. Moreover, if I decide to give them the resources they are asking for, I will need to justify this budget increase to our Board of Directors. And if I decide to cut the projects, how do we decide which initiatives have to be dropped?

COO: While we are on the "strategic issues" topic, there is another concern I wanted to bring up. Once a prominent member of our Board of Directors asked a simple question, "Why did you decide to do the container ship terminal project and postpone the cruise ship terminal one? What made the first project more important than the second?" And we could not provide them with a clear and succinct explanation…. We kind of felt that one was more important than the other, but couldn't—for the lack of the better word—quantify it. We gave them a very generic speech regarding customer satisfaction, growth of local economy, etc., but they weren't that impressed.

Me: Well, in the course of this conversation you touched upon the topics of project prioritization, strategic resource allocation, dropping or killing unwanted projects, and project value. All of these are part of the portfolio management domain.

CEO: What do stocks and bonds have to do with our problems?

Me: Oh, no! You are confusing financial portfolio management with project portfolio management

COO: Never heard of that one!

At first, I didn't pay much attention to this dialogue, thinking that it was just an isolated event. However, in the next several years I was lucky enough to travel around the world doing consulting and training in the project and portfolio management area. As part of my practice, I frequently interacted with C-level people around the world, and to my great surprise, when asked what issues bothered them the most at their companies, the vast majority of the senior managers invariably mentioned the following challenges:

- Lack of resources to complete all of their desired projects
- Projects being delayed, over budget, and not delivering the full scope
- Lack of bottom-line improvements despite all of their project investments

What observations can we draw from this situation? Here is a list of my conclusions:

- The number of ideas flying around any organization is almost always beyond their internal capability (both fiscal and human resource-wise) to handle them.

- Often, the desire of the executives to shove as many projects as possible into the proverbial resource bucket results in projects being under-resourced, which in turn leads to budget and schedule overruns.
- Inability to choose the best projects and, as a result, killing bad ones, causes problems with bottom lines.
- Project portfolio management has provided answers to all of the issues discussed so far, but, unfortunately, due either to the lack of understanding in the executive circles or, at times, a naive belief that a simple installation of a portfolio management software package can address all of the problems faced by modern companies, it has not become as mainstream as, say, project management.

Based on the observations listed here, it became clear that the market needs a new book on project portfolio management. This book was prepared with the following attributes mind:

- It needed to explain the basic concepts of project portfolio management in a simple, comprehensive manner in order to reach the widest possible audience.
- It should focus on both the theory of portfolio management as well as on the real-life application of these concepts so that it can demonstrate the transition from "dry" theory to reality.
- It should contain as many concrete examples as possible in order to demonstrate different facets of project portfolio management.

What I Plan to Do in This Book

This book is not designed to be a comprehensive project portfolio management handbook that would include all possible portfolio management theories, tools, and techniques. What I have attempted to do is focus on practical, simple, and easy-to-implement solutions that can be employed by any company in any part of the world.

A big part of this book focuses on real-life case studies demonstrating how companies around the world, both well-known giants and small, privately held organizations, have successfully developed and implemented their own project portfolio management models and processes.

The book is divided into three sections (see Figure P.1). Section I deals with the theory of project portfolio management and includes

- Chapter 1, Introduction to Project Portfolio Management—A general overview of project portfolio management theory

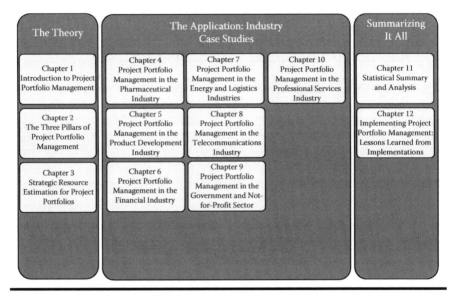

Figure P.1   Table of contents.

- Chapter 2, The Three Pillars of Project Portfolio Management—A detailed description of the concepts of project value, portfolio balance, and strategic alignment
- Chapter 3, Strategic Resource Estimation for Project Portfolios—A discussion of several approaches to enterprise-level resource planning for project portfolios

Section II is dedicated to case studies taken from several key industries:

- Chapter 4, Project Portfolio Management in the Pharmaceutical Industry—Three real-life case studies from the pharmaceutical industry
- Chapter 5, Project Portfolio Management in the Product Development Industry—Seven real-life case studies from the product development industry
- Chapter 6, Project Portfolio Management in the Financial Industry—Four real-life case studies from the banking industry
- Chapter 7, Project Portfolio Management in the Energy and Logistics Industries—Five real-life case studies from the energy sector
- Chapter 8, Project Portfolio Management in the Telecommunications Industry—Four real-life case studies from the telecom industry
- Chapter 9, Project Portfolio Management in the Government and Not-for-Profit Sector—Four real-life case studies from the government sector
- Chapter 10, Project Portfolio Management in the Professional Services Industry—Three real-life case studies from the professional services industry

Finally, Section III of the book concentrates on practical advice for implementing a project portfolio management:

- Chapter 11, Statistical Summary and Analysis—Significant statistics across industries
- Chapter 12, Implementing Project Portfolio Management: Lessons Learned from Implementations—Various ways of deploying project portfolio management and the issues and potential challenges to be aware of when implementing project portfolio management

About the Author

**Jamal Moustafaev, MBA, PMP**, is the president and founder of Thinktank Consulting, Inc., Vancouver, British Columbia, Canada. He is an internationally acclaimed expert in the areas of project/portfolio management, project scoping, process improvement, and corporate training. He has worked for private-sector companies and government organizations in the United States, Canada, Europe, Asia, and the Middle East, including the U.S. Department of Defense, Siemens (Germany), PETRONAS (Malaysia), TeliaSonera (Sweden), and British Petroleum (United Kingdom), to name a few.

Moustafaev authored two other books dedicated to project and portfolio management:

- *Delivering Exceptional Project Results: A Practical Guide to Project Selection, Scoping, Estimation and Management*
- *Project Scope Management: A Practical Guide to Requirements Elicitation, Analysis, Documentation, Validation and Management for All Types of Projects*

Moustafaev is a certified Project Management Professional (PMP*). He holds an MBA in finance and a BBA in finance and management science from Simon Fraser University. In addition to teaching the highly acclaimed "Project Management Essentials" course at the British Columbia Institute of Technology (Vancouver, Canada), Moustafaev also offers several project and portfolio management corporate seminars through his company.

For further information, feel free to contact him:

Jamal Moustafaev, BBA, MBA, PMP

President and CEO

Thinktank Consulting Inc.

E-mail: info@thinktankconsulting.ca

Website and blog: www.thinktankconsulting.ca

THE THEORY

I

Chapter 1

Introduction to Project Portfolio Management

Historical Case Study: Ibaraki Airport

On March 11, 2010, the new Ibaraki (IBR) Airport opened in Tokyo, Japan. The first flight to arrive was an Asiana Airlines Airbus A321 from Incheon International Airport in South Korea. This was the first and last flight that day.

Let us examine this case study from the very "beginning." The airfield was first developed in 1937 under the orders of Emperor Hirohito, and for the next several decades it served as a Japanese Air Force base. Several years after the start of the twenty-first century, the local government decided to convert the military installation into a civil airport.

According to different sources, the cost of the construction project was somewhere between \$220 million and \$230 million. Also, according to multiple publications, the project was completed on time and within budget with all the requested features delivered. Therefore, one could conclude that, from a project management point of view, this project was a complete success.

However, at the time of the project's inception, both of the two major Japanese airlines—All Nippon Airways and Japanese Airlines—notified the local government that they did not intend to use the airport after its completion. These airlines' decisions implied that 90% of the air traffic in Japan would be absent from the airport.

Another issue that was known right from the beginning of the venture was the problematic location of the airport. It was located 96 miles (155 km) from the Shinjuku district of Tokyo. Another problem at the time the airport opened was there were no plans to offer any type of public transportation from or to the airport.

It was estimated that the passengers trying to get to the center of Tokyo would have to spend more than 3.5 hours to reach their intended destination.

Furthermore, the facilities at the IBR Airport were minimal. While the provincial government marketed the airport as a low-cost airline hub, the facilities at the airport were totally insufficient to meet the requirements.

In 2014, there were six local and two international flights to Shanghai and Seoul running from the IBR Airport. This feat was achieved only after a sharp decrease in the landing fees for the airlines. The IBR Airport charged approximately 60% of what the Narita Airport in Tokyo charged the flights for the right to land in its airfield.

As has been mentioned, we cannot really blame the project management aspect for the failure of the project. The team built whatever was required from them on time and within budget. If we cannot hold the project manager responsible for this failure, then who should be accountable?

The answer to that question lies in the project portfolio management (PPM) domain—the art and the science of selecting the best, highest value projects for any given organization. Obviously, the wrong project was selected and implemented by the IBR Airport prefecture in the first place. If the provincial government's strategy has been "we will try our hardest to deliver the biggest bang for the taxpayer's buck," it should have asked the following questions:

■ How will the airport generate revenues for our district if two major Japanese operators, which account for 90% of the country's air traffic, refuse to use our airport even before the construction started?
■ Would any airport located about a 3.5 hours drive from Tokyo attract passengers?
■ Should we consider including some kind of transportation solution to get people to Tokyo?
■ If we are to target the low-cost airlines, should we include the features required by such carriers into the airport design?

Since none of the these questions were asked, the IBR Airport is a symbol of decades of public spending and of vanity projects undertaken by both governments and companies worldwide.

Sounds Comparable to Your Company?

Let me start with a list of top 10 signs that a company you are working for is in dire need of PPM. As we go through the list of signs with appropriate explanations, keep track of what attributes are mentioned in your organization:

1. Project managers and functional managers (department directors and managers) constantly fight over resources. The functional department heads claim that they need their people to fulfill their day-to-day operational obligations,

while the project managers complain that they do not get enough people to finish their projects on time and on budget.

2. Priorities of the projects initiated by the executives constantly changed, resulting in quick resource reassignments. If in January project A was the most important initiative at the company, by June it might be downgraded to number 10 on the list of the important company ventures and may be completely removed from the list.

3. Managers, even at the mid-level, have the authority to unilaterally approve and initiate projects that automatically get added to the company's portfolio of projects.

4. These projects are expected to start as soon as approved by senior managers, regardless of resource availability.

5. There is a chronic shortage of resources at the organization. Employees are constantly complaining about being overworked, while the managers insist that they must roll up their sleeves and work harder.

6. Projects are frequently late and/or over budget and/or do not deliver the full scope promised.

7. Even if the strategic idea is implemented, the company sometimes fails to achieve the expected improvement or fails to receive any value from the project at all.

8. There is significant turnover at the senior management level. A new group of senior executives joins the company, appears cheerful, but at the same time makes vague promises, none of which are realized, and leaves after three to five years.

9. The strategic plan—even if the company has one—is presented as a list of projects, but the cause–effect logic tying those initiatives to the company's mission, goals, and the strategy is absent.

10. The list of company projects is not prioritized. Therefore, it is assumed that all of these initiatives must be started and implemented more or less simultaneously.

If at least five of the attributes match your organization, this book is for you; please read further, learn, and enjoy!

PPM: A Quick Overview

PPM Defined

One of my favorite definitions of PPM states:

> Project portfolio management is the management of the organization's projects so as to maximize the contribution of projects to the overall welfare and success of the enterprise subject to internal and external constraints by maximizing the project value, balancing the portfolio and aligning it with overall company strategy.

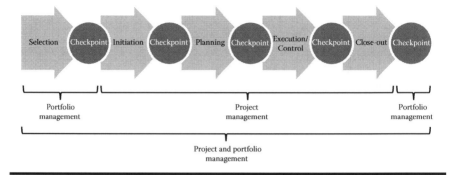

Figure 1.1   Project portfolio management life cycle.

Figure 1.1 demonstrates the proper flow of the project in the PPM life cycle. Initially, someone at the company has a project idea. That person should assess his or her initiative from three aspects—project value, desired portfolio balance, and strategic alignment, and capture all of this information in a business case. For a detailed explanation of value, balance, and strategic alignment, see the "The Three Pillars of PPM" section in this chapter and Chapter 2.

The business case is then submitted to the portfolio selection committee whose mandate is to reevaluate the project according to the approved company's scoring model, portfolio balance requirements, and strategic alignment prerequisites. If the project is approved, the project manager is assigned, and from this point, both project management and PPM run concurrently. The "job" of the project management is to ensure that the project is delivered on time, on budget, and with only minimal defects, while the "responsibility" of PPM is to verify at the end of each stage that the assumptions made about the project value, balance, and strategic fit are still true.

Let us try to visualize this process using a very primitive example. Imagine that someone at a real estate development company decides to build a villa and sell it for profit. Again for the sake of simplicity, let us assume that the company does not care about the balance and strategic alignment, and the only value factor that matters is the return on investment (ROI).

The executive studies the real estate market and comes to a conclusion that the house they plan to build would be worth $100,000. The executive chats with the company architect, who provides them with a very high-level project cost estimate of $50,000. Is this an attractive project? The ROI is calculated as follows:

$$\text{ROI} = \frac{\$100,000 - \$50,000}{\$50,000} = 100\%$$

So, the first checkpoint is passed, the project is approved, and the project manager is assigned. The project manager holds discussions with the project champion, the architect, and several company engineers to create a project charter. Once it is complete, the project cost must be upgraded to $55,000, while the forecasted sale prices remained unchanged. The new ROI then is

$$ROI = \frac{\$100,000 - \$55,000}{\$55,000} = 82\%$$

Again, since 82% is an attractive number, the steering committee approves the project and it moves to the planning stage. Here, the project manager creates the requirements document followed by the villa blueprint and the bill of materials. When the project plan is finished, the cost has to be adjusted to $75,000, mainly because of the unstable grounds where the villa will be located plus higher-than-expected infrastructure expenses. On the other hand, the marketing specialists inform the steering committee that the projected sales price has to be downgraded to $70,000 because of a sharp increase in the interest rates. The ROI now is

$$ROI = \frac{\$70,000 - \$75,000}{\$75,000} = -6.7\%$$

Since the project is no longer attractive by the company standards, it should be stopped until the situation improves. The problem is, as one of the executives states, "Once the locomotive leaves the station, no one even bothers to check on it. It officially becomes a runaway train as soon as it departs!"

One final word of warning regarding PPM: PPM is not to be confused with the following concepts:

- Management of multiple projects—that is, the domain of program management.
- Enterprise project management—that is, a 360° view of the organization's collective efforts.
- Professional services automation—software, no matter how good it is, is not going to choose the right projects for your company.

I have been asked this question in many consulting engagements:

Can we address our project (portfolio) management deficiencies by installing appropriate software?

A short and not very diplomatic answer to this question is an unequivocal "No," and here is why:

Imagine that you can't play a piano. As a matter of fact, you know nothing about music. Will the purchase of the best piano in the world address your inability to play? Probably not …

Another, more technical example:

> *Imagine that you know nothing about accounting to the point that you can't tell the difference between the debit and the credit. Will the installation of the most advanced accounting software on your desktop or laptop instantaneously make an accounting expert out of you?*

Having just a project management or portfolio management software installed on your computers will do nothing to help you with your project-related challenges. As a matter of fact, it is very likely to have an opposite effect as I have witnessed in many organizations. What is likely to happen when people who have a very vague understanding about project management are suddenly forced to fill out endless time sheets and create cumbersome Gantt charts? They will probably fail to appreciate the importance of this and find very creative ways to ignore these tasks.

Now, having said all that, both project management and PPM software implementations after the proper methodologies have been developed and fine-tuned to the company needs can be very helpful. The executives just have to sequence those tasks properly.

The Three Pillars of PPM

One can say that PPM rests on the following three pillars:

1. Projects selected must maximize the value for the company.
2. Projects selected must constitute a balanced portfolio.
3. The final portfolio of projects must be strategically aligned with the company's overall business strategy.

Chapter 2 is dedicated to a detailed analysis of all three concepts; nevertheless, let me share some interesting examples from the experience I gained throughout my consulting engagements.

Initially, many organizational leaders assessed the value of their projects by directly borrowing the portfolio model from the financial industry. In other words, they analyzed the value of their projects based solely on the financial factors, such as net present value (NPV), ROI, internal rate of return (IRR), and many others. However, soon, despite their obvious benefits—companies are in business to make money—these models had two major drawbacks: notorious unreliability of financial forecasts and the fact that the models were ignoring other important factors, such as strategic fit, marketability, resource requirements, risk, etc.

Eventually, the more forward-thinking organizational leader switched to scoring models that included several factors to define and assess the value of their proposed projects. Here are some examples of representative models:

A software product company operating in the e-commerce market developed the following fairly aggressive scoring matrix:

- Product and competitive advantage
- Market attractiveness
- Leverage of core competencies
- Technical feasibility
- Financial reward

On the other hand, a smaller North American university had a more conservative approach to project selection. Its factors included

- Strategic fit
- Resources required
- Technical feasibility
- Financial value
- Riskiness

The second pillar of PPM is the balance of the portfolio. It is usually assessed using 2-D graphs with different values attached to the vertical and horizontal axes. One of the most popular pairs is the project's risk and its financial reward (see Figure 1.2).

In this particular model, the company has projects A and B located in the low-risk, low-reward quadrant, while project C is in the low-risk, high-reward zone. Two smaller ventures, D and E, are positioned in the high-risk, high-reward zone of the graph and, finally, a medium-sized project F is in the high-risk, low-reward zone.

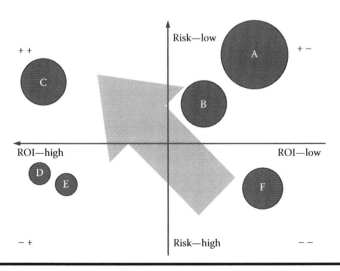

Figure 1.2 Portfolio balance—generic example.

An interesting conversation clearly demonstrating the value of the portfolio balancing took place when I was teaching my Project Portfolio Management Masterclass in the Gulf region. Among other attendees, there were two high-ranking representatives of one of the largest construction companies: an owner (and CEO) and his general manager. The conversation went as follows:

> CEO: This portfolio balancing theory is great but I can hardly imagine how it would apply to my business. We are basically very similar to a professional services company. People come to me and say, "Build me this!" What am I going to reply to them? "Sorry, your project does not fit into our portfolio balance model?"
>
> Me: Well, let me finish the module on balancing the portfolios and we will have a chance to chat about this topic at the end.
>
> CEO: (staring at Burj Khalifa visible through our conference room window) Wait a second! I think I get it! I am fairly old and close to retiring in a couple of years. Your presentation made me think; what kind of legacy am I going to pass on to my son, who will take over our business? Right now our entire portfolio consists of very low-risk, low-reward projects. We basically build shoebox types of buildings with a very low margin of profit. I would like to have that (points to Burj Khalifa) on our company brochures!
>
> GM: Forget about Burj Khalifa, we have conducted some calculations and if we get into HVAC business, our margins will go up from 5% to 25–30%. And if we somehow manage to get into the energy management business, we can raise our profit margins to 50–75%. Too bad we don't have any internal expertise at our company.
>
> CEO: Why don't we hire several specialists in the HVAC and energy management and start a couple of projects from those domains next year? These projects will represent maybe 5% of our total portfolio, but this share will grow with time.

What happened in this conversation? The CEO of the construction company suddenly realized that almost 100% of his projects fell into the low-risk, low-reward category. Concerned with the sustainability of his business model and with the help of his general manager, he decided to shift a small percentage of his projects into the high-risk, high-reward zone, hoping that with experience they would be able to turn them into low-risk, high-reward ventures

The definition of strategic alignment is fairly simple and straightforward: all of your projects must in one form or another assist the implementation of your company's strategy—a very simple statement that at times is very difficult to explain. To do that, let us examine several examples of project alignment and nonalignment.

At one point of time, the executives of Société Bic (commonly referred to just as Bic), a French disposable consumer products company known for its razors, lighters, ballpoint pens, and magnets, made a very interesting decision. The company decided to enter ... the ladies underwear market by designing, producing, and selling, among other things, ladies pantyhose. Needless to say, the company failed miserably with this project since the consumers were unable to see any link between Bic's other products and underwear, because of course there was no link at all.

Although, as the urban saying goes, "hindsight is 20/20," let us nevertheless try to assess this initiative from the strategic alignment perspective. Here is a list of potential questions one could direct at the Bic executives who proposed to add this project to their company's portfolio:

■ We manufacture disposable products made from plastic. What the heck do we know about ladies underwear?
■ All of our production facilities are built based on the injection-molded plastic technology? Where will we get the equipment to manufacture underwear?
■ People, especially females, perceive us as producers of cheap disposable lighters and pen? Would they be interested in purchasing our lingerie products?
■ What about the distribution channels? Retail outlets that trade disposable razors, pens, and lighters usually do not sell underwear. Does this mean we will have to acquire a new group of retail channels?

It is obvious that none of the answers to these questions would have been encouraging had they been asked at the time of project initiation. Indeed, there was little or no alignment between the proposed endeavor and the overall company strategy.

Here is another example that is a bit more subtle, but still very powerful in my opinion. Several years ago, I was hired by a relatively small software company to assess their project and portfolio management practices. After several days of investigation involving interviews with the company's employees and audits of their project management processes and documentation, I jotted the following observations in my notebook:

■ The company consisted of approximately 100 employees roughly divided into two groups: product development (20 people) and professional services (80 people).
■ The product development team was responsible for the continuous development of new versions of the company's products.
■ The professional services guys were the ones responsible for taking the existing platform and deploying it at customer sites.
■ Professional services team charged the customers between $275 and $350 per man-hour, usually generating between $500,000 and $2,000,000 per project in professional services fees.

- The product team, on the other hand, did not generate any revenue.
- While the professional services department was fairly mature from the project management and business analysis perspective, the product development team was a complete mess with an utterly ad hoc approach to their projects.
- As a result, the product team failed time and time again with the delivery of the new product versions.
- The situation got so bad that six out of the eight major customers refused to talk to the company account managers until they fixed their product quality issues.

Further discussions with the product team in attempt to establish the root cause of such a poor performance led to the following discoveries:

- Since the professional services were perceived by the company as "moneymakers" and the product team as "money wasters," all of the best and most experienced resources were always deployed in the professional services department.
- Moreover, if the company was operating at a full capacity and a new customer deployment project came along, instead of hiring additional permanent employees or contractors, the management just cannibalized the product team, again, pulling the best resources and reallocating them to the professional services projects (see Figures 1.3 and 1.4).

Figure 1.3 Product team cannibalization—before.

Figure 1.4   Product team cannibalization—after.

■ Needless to say, both the overall morale and the cohesiveness of the team suffered; add to that lack of any kind of requirements analysis and proper project planning and the overall performance of that team was not that surprising after all.

The subsequent conversation with the company CEO was even more interesting. I did not disclose any of my findings initially to obtain the executive's uninfluenced opinion on the state of company affairs:

> Me: So, let us start at the very beginning. Could you please tell me what the company's mission is? In a perfect world, where do you see your organization in three to five years?
>
> CEO: Well, we intend to become industry's leading provider by being on the cutting edge of innovation and creativity, by supplying the market with the most revolutionary and visionary products.
>
> Me: And who is your competition?
>
> CEO: Companies A, B, C, and D (names several multimillion and even multibillion global brands)
>
> Me: So, you are planning on taking on these giants by having a product development team consisting of twenty inexperienced developers that gets cannibalized in favor of the professional services department every time a new project comes along? How exactly are you planning to accomplish this?

This story serves as one of the best examples showing how the actions of company executives do not align with the overall company strategy. The only thing that remains unclear is whether the strategy was conceived as a set of "sexy" and fashionable words copied from another company's website or, indeed, the executives honestly believed in their mission statement but failed to see how their actions contradicted it.

The discussion of portfolio management in general and strategic alignment in particular would not be complete without the "gut feel project" discussion. I encountered this phrase several years ago when consulting for a very large German company. The conference room was full of division directors—all of them engineers by education—and one of them asked me the following question: "I understand your proposal to establish a selection mechanism for projects, but what about gut feel projects that go against the common sense, but turn out to be ultimate winners? Take Apple and iPhone for example. A company producing computers decided to go into a completely new domain and won."

Yes, Steve Jobs was a visionary, but is it true that his company went into a completely unknown domain? If we examine our smartphones today, what percentage of their functionality is responsible for making and receiving the phone calls? Probably a tiny portion of the overall system and the software installed on it. In reality, modern smartphones are minicomputers with an add-on capability to make phone calls rather than the other way around.

What Steve Jobs was able to predict is that the future of the phones laid in the computer-based technology, and he realized that Apple was very good at designing and building state-of-the-art personal computers. Hence, there was no abandonment of the company's know-how or any other strategic assets when Apple decided to venture into the first iPhone project.

What Happens without Project Portfolio and Proper Resourcing?

There is a multitude of potential problems that await the company without proper PPM processes in place. Initially, lack of portfolio management manifests in terms of reluctance to kill weak project proposals, projects being selected based on politics or emotions, and lack of strategic criteria in the project selection.

What are the immediate results of such an ad hoc approach? There are at least two: too many projects are added to the pipeline and many—if not the majority—of these ventures are of low value to the organization.

These two aspects also have several long-term effects. As the company resources are too thinly spread across multiple initiatives, delivery times tend to increase and the final quality of the products tends to suffer, because the employees are scrambling between multiple ventures, missing deadlines, and making mistakes that become harder to fix as the projects progress from initiation to the close-out stages.

Project failure rate increases either because the initial ideas were of poor value or because—even if they were indeed good ideas—the project teams failed to deliver quality products. As a result, the proverbial "product winners" that every executive craves to see in his company offerings are very hard to come by.

If one can use the sniper analogy, then instead of placing a few well-aimed shots from a high-quality rifle, the company fires multiple blasts from a shotgun hoping that at least some of the pellets will hit the targets.

Another interesting phenomenon that I have observed at many organizations is the accumulation of technical debt that eventually eclipses all of the high-value project work the company can deliver instead.

Let me demonstrate this with a real-life example (see Figure 1.5). I once worked at the IT department of a large financial institution. The executive management of the department had a very interesting approach to their strategic planning: at the beginning of every year, they would examine the previous year's performance statistics and discover that the information technology group has delivered, say, 50 projects. They would go to the strategic planning meeting of the entire company and claim something to the effect of

> Last year we delivered sixty projects. In order to exceed the expectations this year we will accomplish eighty projects!

Obviously, all of the people in the room would be happy with these new commitments, and the new plan would be approved. The interesting aspect of this story is that none of the IT managers even bother to compare the relative complexity of the old versus new projects. Moreover, not one of them even asked a simple question, "How successful were we with the 60 projects we delivered last year?"

They would arrive back at their offices, present the new project list to their employees, and the hard work would commence. The IT team would be assigned the first 20 of the planned 80 of the initiatives (for simplicity, let us assume that 80 projects have been proportionally divided between four quarters). Since they

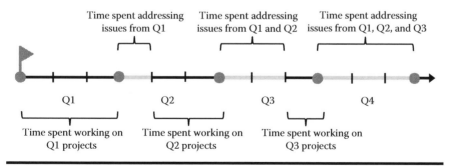

Figure 1.5 The "technical debt" phenomenon.

had trouble delivering 15 projects per quarter and the complexity of the projects usually does not decrease with time, all of the project teams would experience serious issues with the timely delivery of the initiatives assigned to them.

The project managers would tell the executives about the challenges, but they would reply with something along the lines of "Just roll up your sleeves and work harder." At the end of the quarter, the IT management would report that the projects allocated to the quarter have been delivered successfully, although in reality there would be some serious issues, bugs, and deficiencies. What was the response of the business side of the organization? "Great! Here are the next 20 projects! See you at the end of Q2."

What would happen in the second quarter is that the first month of it would be spent addressing the issues left over from the first quarter, which would leave the entire department with two months to deliver the amount of work they could not accomplish in three months in the first place!

The history repeats with the project managers being told to "sweep their problems under the rug" and report to the business side of the organization that everything is working fine. At the end of the second quarter, the business side gives IT an additional 20 projects. The only problem was that the project teams had to spend two out of the three months in the third quarter addressing the issues generated in Q1 and Q2.

When the fourth quarter comes, the department will have absolutely no time to devote to the Q4 projects as its resources were completely invested in correcting the problems generated in Q1, Q2, and Q3.

This particular example has been somewhat fast-forwarded for illustrational purposes. Sometimes, this entire cycle took only a year, but sometimes it stretched to three or four years. However, the end result of not having effective proper portfolio management and strategic planning would always be the same: either a screeching halt to all the company projects or a realization that nothing can be done with the growing technical debt problem.

What Is Happening in the Industry?

In my first book *Delivering Exceptional Project Results* (Moustafaev 2010), I shared the results of Robert Cooper's study (Cooper et al., 2003) regarding the lack of popularity of PPM among various companies:

- 84% of companies neither conduct business cases for their projects nor perform them on select key projects.
- 89% of companies are flying blind with no metrics in place except for financial data.
- 84% of companies are unable to adjust and realign their budgets with their business needs.

The results of this study imply that portfolio management was adopted in between 10 (if you are a pessimist) and 15 (if you are an optimist) organizations as of 2002.

Several years later, in 2012, the Project Management Institute prepared its *2012 PMI Pulse of the Profession*™ (Project Management Institute, 2012) that was dedicated to the topic of PPM. The report was based on an annual global study of more than 1000 projects, programs, and portfolio managers. More than half(!) of the respondents reported frequent use of portfolio management in their organization, an increase of five points from the previous year's survey.

On the one hand, this report shows great improvement from 2002. On the other hand, however, one must take into consideration the audience of the survey. If we survey project, program, and portfolio managers, it is probable that the companies they work for would be more open toward the concepts of project and portfolio management. After all, there are still many organizations without dedicated project managers (let alone program and portfolio managers) on their payrolls.

Nevertheless, these numbers should probably be viewed as a positive trend. Here are some interesting statistics from the PMI study (see also Figure 1.6):

- 62% of projects at organizations that describe themselves as highly effective in portfolio management met or exceeded the expected ROI.
- Of the organizations that consider their portfolio management to be highly effective, 89% claim their executives possess knowledge and understanding of the PPM principles. Compare this statistics with only 25% at the organizations where portfolio management is minimally effective.

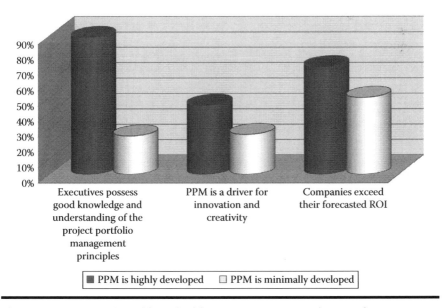

Figure 1.6   Companies with and without PPM—a comparison study.

- When trying to create and foster a culture of innovation, highly effective companies use PPM 45% of the time, as contrasted to only 26% in minimally effective companies.
- Furthermore, according to the study, organizations where managers focus on strategic as well as departmental goals, 70% of projects meet or exceed their forecasted ROI, compared to 50% at organizations where managers neglect strategic alignment.

In addition, the PMI report identified several key drivers for PPM:

- 78% of the respondents mentioned that senior manager receptivity was one of the most important factors.
- 62% said standardized metrics and criteria were important.
- 66% highlighted the importance of competent portfolio governance.
- 59% pointed out the importance of having consistency and logic in organizational strategic objectives.

Conclusions

To summarize our findings so far, consider these lessons from the facts and examples presented in this chapter? Here are the most important ones:

- PPM is important for organizations that want to thrive in the future by being competitive, innovative, and financially driven.
- It is impossible to achieve long-term success by being ineffective with your project selection or hoping that the organization would be able to hire a "visionary CEO," who will be capable of producing one or two brilliant ideas every month.
- Investors are beginning to assess PPM capabilities of a given company before making a decision on whether to purchase their stocks.

Regardless, executives working together with project and portfolio management professionals have additional challenges to address, which include the following:

- PPM is still not widely recognized in the company.
- There is a lack of understanding of PPM.
- Frequently PPM is viewed as something academic, cumbersome, and costly.
- The benefits of PPM may not be obvious to the CXO-level people.
- The task of creating and implementing PPM is frequently delegated to the mid-level managers.

Summary

We started this chapter with an analysis of a project—the IBR Airport construction—that failed from a business perspective but excelled from a project management perspective. The failure to deliver the much-sought value on this project as well as countless other ventures is rooted in the shortcomings on the PPM side.

Later, we looked at the definition of PPM and discussed several examples of portfolio value, balance, and strategic alignment, including a North American product company, a Canadian university, a Saudi construction company, and a software development organization.

We also examined the effect the absence of PPM has on the organizations, including thinly spread resources, longer time-to-market, and poor quality of final products and services.

Finally, we examined two research initiatives—one was conducted in 2002 and another in 2012. The comparison of these studies demonstrates that PPM has made bold strides in the last 10 years, but there is a lot of work to be done, including executive education, spreading portfolio management knowledge, and demonstrating simple and clear examples on how to achieve PPM excellence.

Chapter 2

The Three Pillars of Project Portfolio Management

Introduction

In Chapter 1, we introduced the key pillars of project portfolio management: the value, the balance, and the strategic fit. This chapter provides more details on each one of these domains, shares some relevant examples, and explains how exactly they are achieved.

How to Determine Project Value?

One of the executives who attended my "Project Portfolio Management Masterclass" in London asked an interesting question at the beginning of the workshop, "Could you please explain to us why we need to assess project value? Intuitively I understand that it is important, but once I realize that the project scores 90 points out of 100 (or 7 points out of 50) what do I do with this fact?"

The answer to that question is fairly simple. One of the key assumptions of economic theory is that people are greedy. Not in a negative sense per se, but in a sense that if I pick a random person from a crowd and ask him or her to choose between two piles of money—one with $1000 and another with $2000—the person almost always will pick the second pile.

This same greed can be applied to organizations or, to be more specific, to their executives. When asked—assuming a fixed pool of human and financial resources—whether they would like to implement 100 or 150 in a given year, they will always choose the latter. Unfortunately, just as in the case with individuals and their budget constraints, there are financial and resource constraints associated with project delivery.

The only reasonable answer to this conundrum is to somehow decide which projects will be added to the project pipeline and which ones would be either cancelled or postponed until the required resources become available. But how does one go from looking at a list of 10 endeavors to the point where he or she says, "We will do projects 1, 2, 5, 8, and 9 and will have to cancel projects 3, 4, 6, 7, and 10"?

The only way is rank ordering the project proposals according to some criterion (or criteria) and then inserting a cutoff point at the place where the total resources needed for the projects' implementation equal or exceed the resources available to the organization.

There are two approaches to this methodology: the so-called purely financial approach and the scoring model methodology. Let us examine both of these and consider their advantages and shortcomings.

Financial Models

There are many different approaches to this task, but the most popular ones are the financial method and the scoring model. Let us look at the financial methodology. It implies choosing some type of a financial criterion—be it a net present value (NPV), internal rate of return (IRR), return on investment (ROI), or some other formula—and calculating a value for each project. Once the ROI for each project has been calculated, the projects are ranked according to their ROIs in descending order.

Let us look at an example of how this can be done. Assume we have a company that wants to implement 10 projects and has 200 person-months in its resource pool (roughly 20 people working together for one year including vacation time and allowances for sick days).

Table 2.1 shows a list of projects and their expected ROIs.

Next, the company needs to estimate the efforts required for each project and rank the projects according to their ROIs (see Table 2.2).

It is clear from Table 2.2 that this company can do projects H, E, A, F, C, I, and G, assuming their projections regarding the projects' ROIs and efforts required were correct. Adding project B to the mix will force the company to exceed their effort threshold.

While purely financial models are effective in instilling a sense of discipline and accountability, they all suffer from some inherent problems. One can argue that every financial formula available can be presented in the following form:

$$\text{Financial value} = f(\text{Revenues}/\text{Costs})$$

In other words, any financial value is positively correlated with the project's expected cash inflow and is negatively correlated with the project's cost.

Table 2.1   Sample Project List

Project Name	Estimated ROI (%)
A	21
B	5
C	12
D	3
E	25
F	17
G	8
H	33
I	10
J	4

Table 2.2   Sample Project List—Rank Ordered

Project Name	Estimated ROI (%)	Estimated Total Effort (Man-Months)	Cumulative Total Effort (Man-Months)
H	33	30	30
E	25	40	70
A	21	50	120
F	17	25	145
C	12	15	160
I	10	20	180
G	**8**	**20**	**200**
B	5	10	210
J	4	20	230
D	3	30	260

Bold—this is the last project the company will do; all other projects after that one will be killed or postponed.

Numerous studies confirm that our ability to predict project cost at project inception is somewhere between +300% and –75% (Boehm, 1981) for high-risk industries and between +75% and –25% for familiar endeavors. On the other hand, in many instances, the revenue forecasts are even less accurate with a potential array of values anywhere from –100% to +∞.

Let me prove this controversial point with some examples. When Segway was launched in 2001, it was advertised as the most revolutionary contraption since the invention of the personal computer. The company was forecasting sales of 50,000 units annually. However, they were able to sell only 6000 vehicles.

On the other hand, I seriously doubt that when Steve Jobs conceived the idea for the first iPhone, he expected the sales of this product to exceed the entire annual revenues of Microsoft.

So mathematically speaking, we have a fraction in which the numerator can be predicted with an accuracy of +300%, –75% and the denominator with an accuracy of –100%, +∞. How reliable then is the overall formula?

Another problem with purely financial models is that they ignore factors such as strategic alignment, fit to the existing supply chain, and strategic value of the projects proposed, to list a few.

Scoring Models

Now that we have examined the pros and cons of a purely financial approach to portfolio, let us turn our attention to a more balanced approach—the scoring model.

The essence of the scoring model approach is to have several variables that the executives consider important when assessing the value of their future projects. This is usually done during a project portfolio workshop where the facilitator first explains the theory behind the scoring approach, provides several examples of scoring models developed by other companies, and then asks the executives present to engage in a brainstorming exercise. The essence of this exercise is to generate as many relevant criteria as possible and record them on the whiteboard or a flip chart. These criteria may include, for example,

- Strategic alignment
- Market attractiveness
- Fit to existing supply chain
- Time to break even
- NPV/ROI/IRR
- Product and competitive advantage
- Leverage of core competencies
- Technical feasibility
- Risks

Once the discussion is over, the facilitator gives a red marker to the first person in the room and announces the following rules:

- *Rule #1*: Each participant gets three checkmarks.
- *Rule #2*: Each participant must award all three checkmarks to the attributes listed on the board.
- *Rule #3*: If the participant feels that just one of the attributes is of utmost importance, then he or she awards all three checkmarks to that attribute.
- *Rule #4*: If the participant feels that only two of the attributes are important (e.g., A and B), but attribute A is more important than attribute B, then attribute A gets two checkmarks and attribute B receives one checkmark.
- *Rule #5*: If the participant thinks that any three of the attributes listed are important, then the checkmarks are equally distributed between three attributes.
- *Rule #6*: The number of checkmarks per participant must equal three.

After the first person awarded his or her checkmarks, the red marker is passed on to the next person in the room until all participants have voted on the subject. The facilitator then counts checkmarks awarded to each attribute, and the relative priorities are determined. Here is how it worked in an actual setting: in one of my engagements with a European product company, executives had a list of potentially important project attributes (see Figure 2.1).

Strategic alignment

Market attractiveness

Fit to existing supply chain

Time to break-even

Technical complexity

Product and competitive advantage

Financial value

Leverage of core competencies

Risks

Possible synergies

Competition and IP

Figure 2.1   Scoring variables—before voting.

Strategic alignment ✓✓✓✓✓✓

Market attractiveness ✓✓✓✓

Fit to existing supply chain ✓✓

Time to break-even ✓✓✓

Technical complexity ✓✓✓✓✓✓

Product and competitive advantage ✓

Financial value ✓✓✓✓✓✓

Leverage of core competencies ✓✓✓

Risks ✓✓

Possible synergies ✓✓✓✓

Competition and IP ✓✓✓✓

Figure 2.2   Scoring variables—after voting.

After the "three-point voting" was completed, the whiteboard looked like what is shown in Figure 2.2.

It is clear from the figure that strategic alignment, market attractiveness, technical complexity, financial value, possible synergies, and competition and intellectual property (IP) were deemed to be the most important factors. The next step is to determine boundaries for each criterion. It is important to make these ranges as specific and measurable as possible and avoid terms such as "low," "medium," "high," "weak," and "strong."

See Table 2.3 for the final scoring.

Note that the strategic fit was dependent on the following four factors from the overall company strategy:

1. Create new product families
2. Make products attractive
3. Increase revenue and profitability
4. Increase market share in the new markets

In that particular year, the company's project management office staff calculated that it would have approximately 750 person-months in resources. Note that in this particular example, the organization preferred to measure their resource pool in terms of human reserves available rather than in terms of dollars.

Table 2.3   Sample Scoring Matrix

	1 Point	5 Points	15 Points
Strategic fit	Low Fits one of the criteria	Medium Fits two or three of the criteria	High Fits four or more of the criteria
Possible synergies	Low Cannot combine sales of the proposed product with other product families	Medium Can combine sales of the proposed product with one other product family	High Can combine sales of the proposed product with two or more other product families
Financial value	Minor $0 < NPV < \$1$ million	Medium $\$1$ million $< NPV < \$5$ million	Major $NPV > \$5$ million
Technical complexity	Very difficult Significant external expertise is required	Somewhat difficult Will need some external expertise	Easy Can be implemented by internal employees
Market attract	Low Less than 10 requests	Medium Between 11 and 30 requests	Major More than 30 requests
Competition and IP	High Many competitors Weak IP protection	Medium three or four competitors Normal IP protection	Medium zero or two competitors Strong IP protection

Furthermore, the product company had seven projects to prioritize. Table 2.4 shows the projects, their scores for each category, total scores, and the resource requirements.

After all of the scores are calculated, all that as left is to resort the table according to the total project scores in descending order as well as adding a Cumulative Resources column. As can be seen from Table 2.3, considering the constraint of 750 person-months, the company can do only projects O, M, R, S, and N. Projects P and T would have to be either dropped or postponed until the next year (see Table 2.5).

Table 2.4 Project List before Ranking

Project Name	Strategic Fit	Possible Synergies	Financial Value	Technical Complexity	Market Attraction	Competition and IP	Total	Resources
M	15	5	1	5	5	15	**46**	150
N	5	5	1	5	5	5	**26**	250
O	1	15	15	5	15	5	**56**	200
P	5	5	1	5	5	5	**26**	170
R	5	15	1	15	5	1	**42**	50
S	1	5	5	5	15	5	**36**	100
T	1	1	5	5	5	5	**22**	150

Table 2.5   Project List—after Ranking

Project Name	Total	Resources	Cumulative Resources
O	56	200	200
M	46	150	350
R	42	50	400
S	36	100	500
N	**26**	**250**	**750**
P	26	170	920
T	22	150	1070

Bold—this is the last project the company will do; all other projects after that one will be killed or postponed.

How to Balance Portfolios?

Portfolio balance is important for several reasons. While assessing the value of the projects proposed, it is easy to lose the sight of the "big picture" and suddenly end up in a situation where the company has a large number of small, relatively meaningless initiatives and few significant breakthrough endeavors.

Furthermore, it is also possible that specific areas of the business—especially the departments that are perceived to be the "money makers"—receive a disproportionate number of new projects. Several experienced executives also mentioned in conversations with me their desire not to keep "all of their eggs in one basket" when attempting to balance their portfolios.

Chapter 1 discussed the Gulf construction company example, and we will refer to it in this section. This company has been involved in a multitude of construction projects described by the company owner as "four walls and a roof." These projects were relatively simple and straightforward, involved little risk, and consequently carried small profit margins. In other words, the entire project portfolio of the company resembles that in Figure 2.3.

After a brief discussion that took place during my "Project Portfolio Management Masterclass," the owner of the company together with his general manager decided to shift some of their resources into riskier domains—heating, ventilation, and air conditioning (HVAC) systems, as well as energy-efficient buildings. Their plan was that in the immediate future, the company portfolio will still consist of a large percentage of low-risk, low-reward projects, but yet a small portion of the endeavors would fall into the high-risk, high-reward category (see Figure 2.4).

However, they hoped, as the company's professionals obtained more experience in the HVAC and energy management domains, the high-risk, high-reward

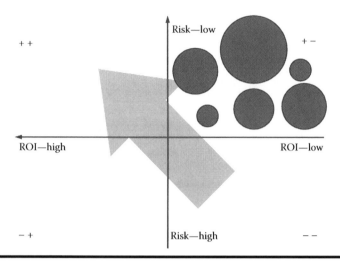

Figure 2.3   Portfolio balance: construction company—before.

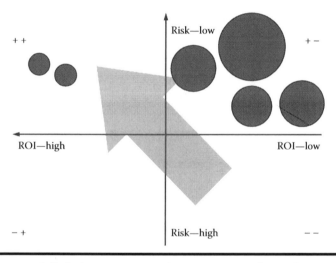

Figure 2.4   Portfolio balance: construction company—after—short term.

projects will gradually rise to the top part of the chart and become low-risk, high-reward endeavors (see Figure 2.5).

Here is a slightly different example of another portfolio balancing challenge. I was part of a consulting project with a company that produced bearings. It had been in this business for several decades producing a variety of highest-quality models, when its sales department initiated a mini-revolution of sorts. The chief sales officer approached the company's president one day and said, "Listen, I know that we pride ourselves on our bearings, but we constantly get calls from our customers

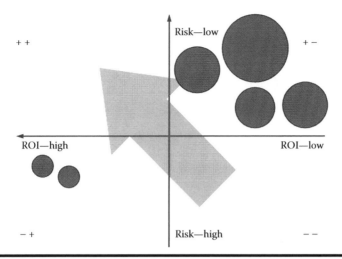

Figure 2.5   Portfolio balance: construction company—after—long term.

asking for stuff like lubricants, sealants, and electronic components. They are all complementary products to our bearings portfolio of products. And we have to reply to these inquiries by claiming that we only focus on one product. Do you realize how much revenue we are losing every year because of that?"

This discussion led the CEO to assign the research and development (R&D) department the task of developing new products in the three new categories: lubricants, sealants, and electronic components. Chapter 5 discusses this company's project portfolio model. Figures 2.6 and 2.7 show "before" and "after" snapshots of its R&D project portfolios.

As can be seen from the charts, this company's R&D portfolio shifted from 100% investments in bearings and 0% in lubricants, sealants, and electronic components to only 10% in bearings and 33% in lubricants, sealants, and electronic components.

In general, the so-called risk–reward diagram seems to be the most popular bubble chart across industries with approximately 44% of companies preferring it over other models (Cooper et al., 2002).

Let us examine the "risk–reward" diagram in a bit more detail. Usually, along the vertical axis, we have a risk variable: it could be an overall risk, a commercial risk, a technical risk, or even a mathematical inverse of risk—the probability of success. Along the horizontal axis, we position a financial variable: ROI, IRR, or NPV. Traditionally, the scale on the horizontal axis increases from left to right (see Figure 2.8).

The two axes divide the entire plane into four quadrants called, respectively, "bread and butter," "pearls," "oysters," and "white elephants." Let us examine each one in more detail.

Bread and butter projects are low-risk, low-return endeavors. They include the traditional revenue sources for the organization (e.g., "four walls and a roof" initiatives in the earlier example) or maintenance projects (e.g., replacement of old servers).

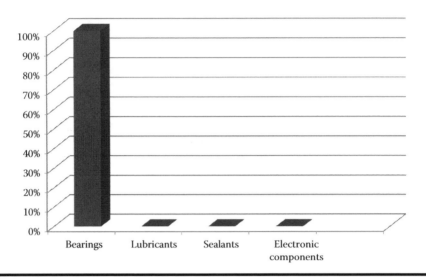

Figure 2.6   Portfolio balance—before.

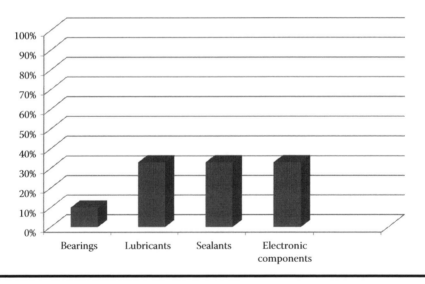

Figure 2.7   Portfolio balance—after.

In most cases, they are a reliable source of income, a proverbial "cash cow," and their representation in the company's portfolio depends heavily on the industry's conservativeness. In other words, one should expect to encounter a much higher percentage of such projects in the banking industry rather than in the video gaming sector.

"Pearls" on the other hand—just like the names suggests—are unequivocal winners that every organization wants to have in its portfolio. These projects possess

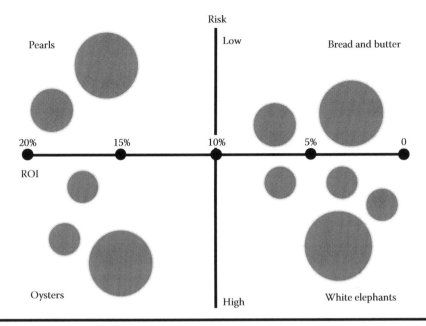

Figure 2.8 Sample risk–reward diagram. *Legend:* **Circle radius = project size.**

a rare combination of high reward and low risk that makes them attractive to every executive, whether he or she is from the ultraconservative insurance industry or the freewheeling smartphone apps domain.

"Oysters" are the high-risk, high-reward projects. Projects involving new products, new product families, or unknown technologies usually fall into this category. They are somewhat dangerous because neither technical nor commercial success is guaranteed, but unless some of the company resources are invested in such initiatives, it is unlikely that the company would be able to compete in the marketplace in a year or two. As in the case of "bread and butter" projects, it is difficult to determine the percentage of such projects in a company's portfolio.

Finally, consider the "white elephant" projects or the initiatives that are high risk but provide low ROI. These projects are unequivocal candidates for either an immediate corrective action or an outright cancellation unless there are sound reasons to keep them in the organizational portfolio. For example, typically most of the enterprise resource planning (ERP) system implementations inevitably fall into this category. On the one hand, it is difficult to create a financial model that would justify their profitability, and, on the other hand, they tend to be large, complex, and, as a result, risky.

In this context, it is also interesting to learn a bit more about the roots of the term "white elephant." It originated from the kings of Thailand: whenever they wanted to punish one of their subjects, they would give them an elephant. The animal required so many resources in terms of food and care that the receiver of such a gift almost inevitably went bankrupt after a fairly short period of time. Needless to say, killing an animal or even letting it escape into the wild was not a viable option because that would have been considered an insult to the king.

Finally, despite the fact that the vast majority of companies prefer to measure their portfolio balance via the risk–reward bubble charts (Cooper et al., 2002), there is a multitude of other alternatives available, which include

- Ease vs. attractiveness
- Strength vs. attractiveness
- Cost vs. timing
- Strategic fit vs. financial benefit (reward)
- Cost vs. financial benefit (reward)

In addition, bar charts can be used instead of bubble charts to compare projects based on markets, market segments, product lines (see the bearing manufacturer example discussed earlier), and technology or platform types.

What Is Strategic Alignment?

An easy way to explain the point of strategic alignment is to ask the reader to imagine the following situation.

Imagine that someone walks into the office of your company's CEO and asks him to produce a list of the current projects at the company. Pretend also that your CEO is actually capable of producing such a list (not typical at many companies). Afterward, the visitor starts to point in an absolutely random fashion at the projects on the list and asks the same two questions over and over:

Why are you doing this project?

and

How is this initiative related to your company strategy?

First, let us examine a couple of simplistic examples. Let us pretend that one of the projects on the list was "Open a Sales Office in Brussels." If the CEO explains this project by pointing to the company strategy of expanding its presence in Europe, then this project is aligned with the overall company strategy.

However, if the company strategy states that it will be aggressively expanding its presence in the Asian markets and does not mention the European region at all, there is a good chance that this project is not aligned with the overall strategy.

Now that we have examined a simple example, let us review a real-life "strategy-to-project" tree developed recently by a European pharmaceutical company (see Figure 2.9).

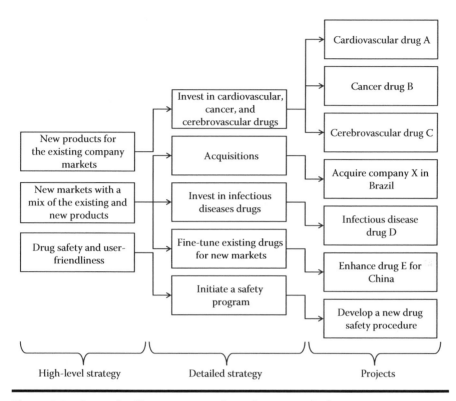

Figure 2.9 Strategic alignment example—pharmaceutical company.

The Board of Directors approved that the overall high-level strategy of the company was in 2013. It consisted of three strategic components:

1. New products for existing company markets—development of new types of drugs for the U.S. and western European markets
2. New markets with a mix of the existing and new products—creating new drugs specific for new markets in Brazil, China, and Russia; also fine-tuning existing company drugs to fit the requirements of these markets
3. Drug safety and user-friendliness—giving special attention to the safety and user-friendliness of both new and old drugs produced by the company

The first macro strategy resulted in creating three new research areas in the company's R&D department:

1. Cardiovascular drugs
2. Cancer drugs
3. Cerebrovascular drugs

These three areas resulted in three distinct projects:

1. Cardiovascular Drug A development project
2. Cancer Drug B development project
3. Cerebrovascular Drug C development project

The second high-level strategy led to three different micro strategies:

1. Acquire new businesses
2. Invest in infectious diseases drugs
3. Fine-tune existing drugs for new markets

The company executives then initiated the following projects:

■ Acquire company X in Brazil
■ Develop infectious disease Drug D
■ Enhance Drug E for China

The drug safety initiative resulted in a project to develop a new drug safety procedure for the company.

Figure 2.9 represents a perfect—albeit somewhat simple—example of a project-to-strategy linkage. Any project selected from the right side of the diagram can be easily linked to the high-level company strategy.

On a more general note: If you have identified a project in your company portfolio that cannot be traced back to the overall organizational strategy, you are most likely looking at an endeavor that is wasting organizational resources.

The three strategic alignment methodologies used in the modern project portfolio management domain are

1. Top-down approach
2. Bottom-up approach
3. Combined top-down, bottom-up approach

Top-Down Approach

The top-down methodology implies that the senior managers decide both on the strategic allocation and the specific projects to be implemented without any participation from the mid-level and junior members of the company. The most popular methods in the top-down approach are the product road map and the strategic buckets model.

The product road map implies a series of product or platform developments on a timescale. In other words, we are standing at an imaginary starting point today, peering into the future and attempting to predict how our products or services will develop. Figure 2.10 shows a simplified fictional example of a domestic washer road map.

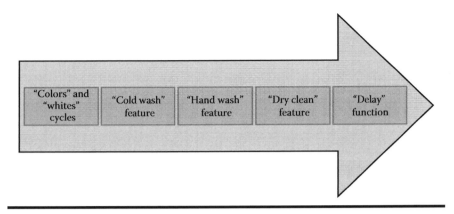

Figure 2.10   Sample road map—washer.

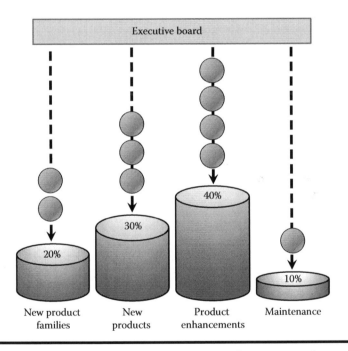

Figure 2.11   Sample strategic buckets model—top-down approach.

Initially, this company was planning to develop a washer with the "colors" (warm water) and "whites" (hot water) cycles. The second release was supposed to include a cold wash cycle, the third one, the "hand wash" capability, the fourth one, the "dry clean" feature, and finally, the fifth one, the "wash delay" function.

The "strategic buckets" model implies two steps (see Figure 2.11):

1. The executives decide on how to split their project resources (both financial and human) across various markets, technologies, or project types.
2. The executives pick the projects they prefer and deposit them into the strategic buckets until all the buckets are full.

The top-down method has both distinct advantages and shortcomings. The pros of this methodology are as follows:

- Executives are responsible for extending the projects from the top to the bottom of the organization.
- Optimistically, the projects flow from the high-level strategic goals of the company.
- Projects tend to be more strategic and long term in nature.

But there are several potential issues with this approach:

- The executives must have a clear and simple business strategy.
- It requires the executives to be precise about how and where they want to spend their financial and human resources.
- In the case of product road maps, the executives must become "fortune tellers" of sort; something that is very difficult to do in the business world.

Bottom-Up Approach

The bottom-up approach is popular in functional organizations including government agencies, banks, insurance companies, and traffic authorities, to name a few. The essence of this approach is that every department submits a list of the projects it needs to accomplish in the next year. All of the project proposals are gathered by the executive board and channeled through the scoring matrix. The projects that receive the maximum number of points—subject to resource constraints—are added to the active company portfolio (see Figure 2.12).

The strengths of this model include its simplicity and straightforwardness. In addition, in the long run, it decreases the number of "frivolous" project requests; what is the point of submitting meaningless ideas if they get cut year after year?

However, this model has two major flaws. First, since the proposals are generated from individual departments, the entire process could lead to an improper portfolio balance. For example, the organization may end up with a lot of high-scoring "new product" projects and few system maintenance initiatives.

Second, since the entire idea generation initiative is given to the departments, it is likely that the final portfolio will be dangerously skewed toward tactical projects serving the needs of the respective organizational divisions, while the number of breakthrough strategic initiatives, usually instigated by the senior executives, will be dangerously low.

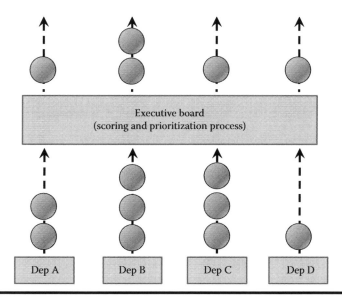

Figure 2.12   Sample strategic buckets model—bottom-up approach.

Top-Down, Bottom-Up Approach

The top-down bottom-up approach is the most popular, as will be demonstrated in subsequent chapters. It combines the advantages of both the other two methodologies, while eliminating almost all of their shortcomings.

The top-down, bottom-up approach consists of the following simple steps (see Figure 2.13):

- The executives define the company strategy, the scoring model, and the desired portfolio balance.
- Anyone at the company, including executives, can submit his or her project proposal to the executive board.
- Projects are scored and distributed to the relevant buckets until all of the buckets are full.

Note: For a detailed description of this process, please see the next section "How It All Works in Real Life."

The benefits of this methodology are as follows:

- Company strategy is well defined.
- Scoring model is employed to assess the value of each project.
- Portfolio balance buckets are utilized to ensure proper distribution of company initiatives.
- All members of the organization—including frontline workers and executives—can contribute their project ideas.

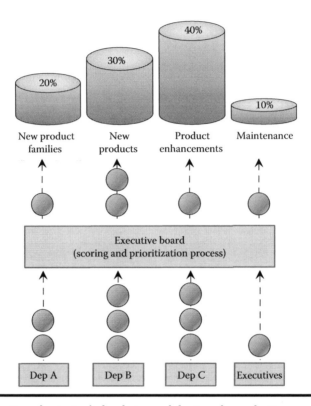

Figure 2.13   Sample strategic buckets model—top-down, bottom-up approach.

How It All Works in Real Life

We discussed the project scoring process earlier in this chapter. This section considers another, more complicated example where the company is trying to prioritize its projects based on both the project scores and from the strategic alignment point of view.

The company in question, a European telecommunications player, had the following list of project proposals at the beginning of the fiscal year (see Table 2.6). The first column contains the titles of the projects, the second one their scores according to the company's valuation model, the third one type of the project (breakthrough, enhancement or maintenance), and finally the fourth one project cost in millions of dollars.

In addition, the company has allocated a total of $40 million for the new ventures. The executives decided to split these resources in the following manner (see also Figure 2.14):

- Breakthrough projects—$8 million
- Enhancement projects—$20 million
- Maintenance projects—$12 million

Table 2.6   Sample Project List

Project Name	Project Score	Project Type	Project Cost
A	50	B	4
B	45	B	2
C	41	E	8
D	40	B	1
E	36	B	1
F	35	M	4
G	31	B	5
H	27	E	4
I	23	E	2
J	30	E	2
K	26	M	4
L	22	E	7
M	18	M	2
N	14	M	1
O	5	M	1

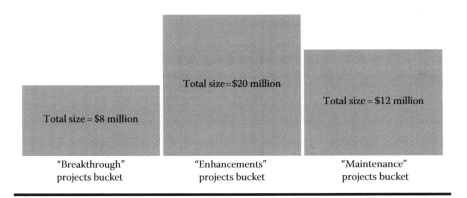

Figure 2.14   Strategic buckets project allocation—step 1.

This breakdown was determined purely based on the executives "guesstimates" regarding how their company should develop in the next several years. Based on my experience, I can confirm that this breakdown was fairly ambitious for a mobile services provider, with almost 20% of the total portfolio budget allocated to new breakthrough endeavors. This ambitious approach could be partially explained by the fact that the organization once dominant on the local market—with a market share of more than 60%—lost some ground in the last couple of years. As a result, the Board of Directors mandated the executives to regain the company's previous dominance in the market.

The scoring criteria used in this exercise were

- Financial (ROI)
- Competitive advantage
- Improves customer satisfaction
- Innovativeness
- Strategic fit
- Time to market

where strategic fit included

- Improve customer loyalty
- Increase market share
- Develop regions
- Improve public image

In addition, the popular top-down, bottom-up model was used for strategic alignment purposes. The portfolio balance was partially addressed by the choice of the strategic buckets. In addition, the managers decided to monitor the portfolio balance using ROI vs. risk bubble charts throughout the life of the current project portfolio.

Also, it is worthwhile to mention that with the selected scoring model, it was expected that the "breakthrough" projects would score on average higher than the "enhancement" initiatives, while the "maintenance" ventures would probably end up near the bottom of the portfolio.

The process starts at the top of the table with project A as the highest ranking project in the portfolio (a score of 50 points). Since this is a "breakthrough" project, it is deposited into the first bucket. Project B (a score of 45) is also placed into the "breakthrough" bucket. The sorting process continues with project C going into the "enhancement" bucket and projects D and E going to the "breakthrough" bucket (see Figure 2.15). Then, it is interesting as the "breakthrough" bucket is now full. This implies that even if we encounter any more "breakthrough" projects in our table (project G), they will need to be automatically rejected.

The process continues with H, I, J (all "enhancements"), and project K ("maintenance") being deposited into respective buckets (see Figure 2.16). At this point in

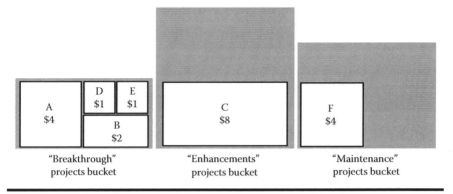

Figure 2.15 Strategic buckets project allocation—step 2.

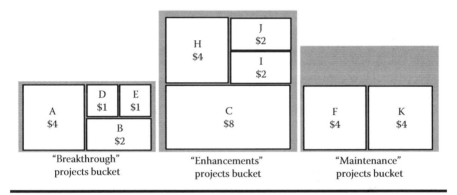

Figure 2.16 Strategic buckets project allocation—step 3.

the process, we encounter yet another constraint: the "enhancements" bucket is also full. This, again, implies that even if we encounter more "enhancement" projects that have a higher score than the remaining "maintenance" initiative, they will have to be discarded or deferred. As a result, project candidate L cannot be considered (see Table 2.7).

Finally, the remaining highest scoring proposals M, N, and O are deposited into the last bucket, and the portfolio prioritization and alignment exercise is complete (see Figure 2.17 and Table 2.7).

The final point to be made is that, frequently, after such an exercise, executives discover that the scoring model is flawed as it prevented projects that had to be implemented from ending up in one of the buckets.

There are two ways to address this issue. One is to recalibrate the entire scoring model, for example, by adding or replacing one or two of the variables. In one of my consulting experiences, the senior managers decided to add a "competitive advantage" variable to the overall model.

Table 2.7   Final Rankings

Project Name	Project Score	Project Type	Project Cost
A	50	B	4
B	45	B	2
C	41	E	8
D	40	B	1
E	36	B	1
F	35	M	4
G	31	B	5
H	27	E	4
I	23	E	2
J	30	E	2
K	26	M	4
L	22	E	7
M	18	M	2
N	14	M	1
O	5	M	1

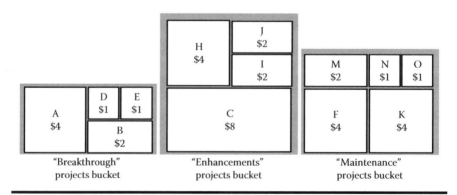

Figure 2.17   Strategic buckets project allocation—step 4.

Joker Project Concept

Another popular way to circumvent this problem is to introduce the "joker project" concept. This methodology places the project candidate at the top of the rank-ordered list of project proposals even if it scores very low in the current company's scoring model.

Sometimes, the portfolio steering committee has a proposal that scores low across almost all of the scoring criteria, and yet the key decision makers feel that it is an important and valuable initiative that can become the next breakthrough project that would generate millions, if not billions, of dollars for their company.

Let me use a somewhat "fantasy" scenario to illustrate this concept. The work on the first iPhone started at Apple in 2004. I am unsure whether Apple used a portfolio scoring model to assess its project ideas. If it had one, I do not know what variables Apple used. So, let us make two not-too-far-fetched assumptions:

1. Apple does use a portfolio scoring model
2. Apple's scoring algorithm is not too different from other software product companies

If we continue our logical line of thinking, then it is safe to expand the second point and assume that its imaginary scoring model includes parameters such as financial value, competitive advantage, technical risk, commercialization risk, technical feasibility, and time to market, to name a few.

Let us pretend that we are in the same room with Steve Jobs where he just proposed to embark on creating something called a "smart phone." What would the assessment of this proposal look like in 2004?

- *Financial value*: Very inconclusive. Very high project costs and unpredictable revenues (see the "Commercialization Risk" discussion). Verdict: either 1 point out of 10 or 5 points out of 10 if we are to use the traditional scoring model.
- *Competitive advantage*: Again, inconclusive. Yes, we will be the first ones on the market with the touch screen model, but there were several attempts before us; some of them failed, and some became very popular. Verdict: 5 points out of 10.
- *Technical risk*: Very high. Yes, we know about producing cool computers, but this project takes us into an unfamiliar domain. Also, new touch screen and apps technologies have never been used before. Verdict: 1 out of 10 points.
- *Commercialization risk*: Very high. Will people like this new product? Will they be willing to pay $600 per unit? Will the mobile companies collaborate with us and create attractive data plans for iPhone users? Verdict: either 1 point out of 10 or 5 points out of 10.
- *Time to market*: Very unpredictable because of the technical challenges to overcome, which probably are longer rather than shorter. Verdict: 1 point out of 10.

So, what is the total value of the proposed project? Either 9 or 17 points (out of 50) depending on whether one is pessimistic or optimistic. Either way it is a low score, and the project is a candidate for cancellation. And yet—we are still in our fantasy scenario—Steve Jobs overrules the decision of the steering committee and insists on undertaking this project. I do not have to remind you how this little venture ended: in 2012, iPhone sales alone exceeded the entire sales of Microsoft, including revenues from Windows software, Office suite, Xbox, Bing, and Windows Phone!

Another, less "glamorous" example of a need for a joker project was one I encountered at a small North American university. The executive board recommended a project proposal to upgrade the university's student information systems that included modules such as student information management, online learning, assessment development and analysis, curriculum mapping, special education, and finance and human resources. At the time of this conversation, the "university ERP platform," as it was often referred to by the university's employees, had not been upgraded for nine years, and it had been slowly crumbling, causing more and more problems over the years.

The university's scoring model looked like this:

- Strategic fit (included components like attracting more local and international students, improving the university's reputation locally and internationally, providing the best possible mix of services and benefits to students and employees, and increasing the social value of programs and initiatives undertaken)
- Resources required (the less resources are need, the more attractive is the project)
- Technical feasibility (the more external resources are required, the less attractive is the project)
- Financial value (either revenue generation or cost savings)
- Riskiness (included reputation, regulatory, financial, or operational disruptions risks)

How did the proposed project score in each one of these categories?

- Strategic fit—Very low, no impact whatsoever on attracting students, improving reputation, or increasing the social value, although one can argue that it helps to improve the mix of services and benefits
- Resources required—Very low, as the project in question was expected to be the largest ever the university has undertaken
- Technical feasibility—Very low, since most of the resources on that project had to be external
- Riskiness—Again, very low because the project would have exposed the school to all of the risks listed, including reputation, regulatory, financial, or operational disruptions

Interestingly enough, despite its alarmingly low score, the project received an approval from the executive committee for one simple reason: the university would have ceased to function if this issue remained unattended for another year or two.

What can be said at the end of this section? The "joker power" should be used sparingly and only for the projects that are of extreme importance for the organization. They usually fall into one of two categories: business continuity or the next great breakthrough project that will change the fortunes of the company. In either case, the responsibility of the project's success or failure will rest upon the executive committee.

Summary

We started this chapter by discussing two different approaches to determining project value: the purely financial approach and the scoring model methodology. We examined their pros and cons and concluded that in most cases the scoring model method is more comprehensive than the financial one.

Also, we examined the importance of portfolio balancing and looked at several examples involving various balancing approaches including "by product type" and "risk versus reward."

We examined the concept of strategic portfolio alignment and studied three main approaches: top-down, bottom-up, and the combined "top-down, bottom-up" method, the most popular in the industry. We looked at more examples of such approaches and their advantages and disadvantages.

Finally, we demonstrated the entire process of portfolio prioritization, balancing, and strategic alignment using the scoring model and top-down, bottom-up approaches. The last section of the chapter discussed the "joker project" concept that sometimes can allow a low scoring but nevertheless promising project to rise to the top of the rank-ordered proposal list.

Chapter 3

Strategic Resource Estimation for Project Portfolios

Introduction

In all of the project portfolio management prioritization and resourcing examples we have considered so far in this book (see section "How It All Works in Real Life" in Chapter 2), we have made one very unrealistic assumption: that the project cost—either in terms of dollars or in terms of resources required—has been known with a high precision rate right from project inception (see Table 3.1). Unfortunately, nothing can be further from the truth in the exciting and highly unpredictable world of program and project management (PPM).

Chapter 2 mentions estimates can vary significantly at the inception when compared to the final cost, duration, or resource requirements. One of the most important things that the executives simply must understand before embarking on any portfolio management activities is that uncertainty lies in the heart of the project estimation process. Before the start of the program or project, there are myriad questions that remain unanswered until at least the planning stage starts:

- How many features (high-level requirements) will be included in the project scope?
- Will the stakeholders decide to include Feature A in the scope?
- If Feature A is included in the scope, would the stakeholders prefer a simple, average, or sophisticated version of the feature?
- If the team implements the simple version of the requirement, will the stakeholders change their mind and ask for a sophisticated version?

Table 3.1   **Sample Project List**

Project Name	Project Score	Project Type	Project Cost
A	50	B	4
B	45	B	2
C	41	E	8
D	40	B	1
E	36	B	1
F	35	M	4
G	31	B	5
H	27	E	4
I	23	E	2
J	30	E	2
K	26	M	4
L	22	E	7
M	18	M	2
N	14	M	1
O	5	M	1

- How will Feature A be designed and built?
- How long will it take to address all the potential risks, constraints, alternatives, and exceptions associated with building Feature A?

Figure 3.1 demonstrates the concept of uncertainty from two distinct points of view: the Project Management Institute's (PMI®) benchmarks for all types of projects and historical data from the software development industry.

A Guide to the Project Management Body of Knowledge (PMBOK®) recommends using the following estimate ranges for projects:

- +75/–25% for the initiation stage (PMI, 2012)
- +30/–15% for the planning stage
- +10/–5% for the execution and control stages

In other words, if the actual resource requirement for the project turns out to be 1000 person-months, it is absolutely acceptable for the project manager to provide the executive team with an initial estimate ranging from 750 to 1750 person-months.

A study of IT and software companies conducted by Barry Boehm (the guru of software estimation) demonstrated that the funnel for software development

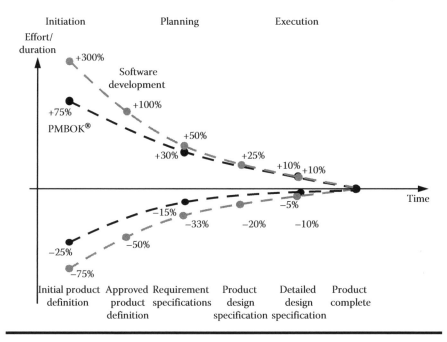

Figure 3.1 Cone of uncertainty.

projects was even wider (Boehm, 1981). Among other things, it showed that even software development companies that excelled were in the +300% to –75% accuracy range when estimating early in the project life.

In the light of these facts, we should now ask the most important questions:

> *How are we supposed to deal with estimate uncertainty when running project portfolio planning exercises?*
> *Is there a way to include certainty and probability into the PPM resourcing exercise?*

In the next sections of this chapter, I will propose several innovative ways of dealing with these types of issues.

Improving Your Estimate Accuracy with Wideband Delphi and PERT*

Wideband Delphi

This methodology is not the most reliable when compared with the Program Evaluation and Review Technique (PERT), but it is faster and still better than single-number estimates.

---

* Adapted from Moustafaev (2014).

The Wideband Delphi was created in the early 1940s by the Rand Corporation at the time of the creation of the atomic bomb (the Manhattan Project). Barry Boehm later refined this methodology for technology project purposes (Boehm, 1981).

The "industrial strength" version of the Wideband Delphi technique consists of the following steps (*Note*: the steps have been modified for whole projects rather than project tasks):

1. The coordinator presents each estimator with a detailed project description available at the time of discussion and an estimation form.
2. Estimators discuss project scope, risks, and potential complexity issues with each other (but not the estimate itself).
3. Estimators fill out forms anonymously (important).
4. The coordinator prepares a summary of the estimates on an iteration form (similar to the estimation form).
5. The coordinator has estimators discuss variation in estimates examining range, average, and extreme values.
6. Estimators fill out forms again, anonymously, and repeat steps 4–6 as many times as needed.

Notice several peculiarities in this process. First, estimators are free to discuss the project and the complexity associated with implementing it, but no one is allowed to say something along the lines of

I think Project A should take no more than 120 days
or
Project B will cost us $150,000

Furthermore, the estimation forms must be filled out anonymously. These steps are undertaken to ensure that vocal and strongly opinionated people on the team do not influence the quiet and shy team members, who may, and frequently do, possess more accurate information.

Then, after collecting all the estimation forms, the project manager writes all of the estimates on a whiteboard:

5 months
4 months
6 months
4 months
6 months
5 months
20 months
4 months
5 months

We can see almost all of the estimates are fairly similar and range between four and six months, with one exception—the 20-month estimate provided by one of the project team members. Rather than asking whomever came up with the larger estimate to stand up and explain his or her reasoning in front of the rest of the team, the project manager is expected to ask something along the following lines, "And why do you think some of us may believe that this project may take 20 months?"

It could happen that most of the team members in the room may think that the delivery of the proposed project should take no more than four months. And yet, there is one member of the team—a marketing expert—who knows for a fact that a new advertising campaign has to be created and implemented in order for the initiative to succeed, thus increasing the project duration to 20 months. The potential problem in this situation is that the only person who knows the true duration (of effort) of the project may be overwhelmed by the opinions voiced by more vocal team members.

Thus, the appropriate attitude is that no one knows the right answer, and the team members are not allowed to discuss the actual durations or efforts of the projects among themselves.

Wideband Delphi "Light"

Wideband Delphi "light," on the other hand, disperses all the formalities typically unnecessary on the majority of projects. For example, estimation forms can be replaced with small pieces of paper, and estimates are recorded on the whiteboard or a flip chart.

Program Evaluation and Review Technique

PERT was invented by Booz Allen Hamilton, Inc., under contract to the DOD's U.S. Navy Special Projects Office in 1958. The endeavor it was working on was the Polaris mobile submarine-launched ballistic missile project.

Let us pretend that we are working on a fairly simple and straightforward portfolio consisting of the following projects:

- Project A
- Project B
- Project C
- Project D
- Project E
- Project F
- Project G
- Project H

Table 3.2 Sample Portfolio PERT Estimation

Project Name	Opt (in '000 $)	ML (in '000 $)	Pess (in '000 $)	PERT Mean (in '000 $)	PERT St. Dev (in '000 $)	PERT Var
A	100	130	150	128.33	8.33	69.44
B	50	55	60	55.00	1.67	2.78
C	50	100	200	108.33	25.00	625.00
D	120	140	160	140.00	6.67	44.44
E	30	45	50	43.33	3.33	11.11
F	45	60	70	59.17	4.17	17.36
G	200	250	300	250.00	16.67	277.78
H	20	25	40	26.67	3.33	11.11
				810.83	**32.54**	1059.03

In the next step, we conduct one or several Wideband Delphi exercises with the entire executive team and generate the optimistic, most likely, and pessimistic duration estimates for each project with the assistance of the organization's project managers (see "Opt," "ML," and "Pess" columns in Table 3.2).

People frequently ask, "While we understand the concept of most likely estimates, how does one come up with optimistic and pessimistic ones?" The suggestion is to think of an optimistic estimate in the following manner:

If everything that can go right will go right on this project, how long will it take (how much will it cost, how many person-days will it require)?

Coming up with a pessimistic estimate, on the other hand, implies answering the following question:

If everything that can go wrong will go wrong on this project, how long will it take (how much will it cost, how many person-days will it require)?

The mean duration, the standard deviation, and the variance for each project are calculated based on the following formulas (see "PERT Mean," "PERT St. Dev," and "PERT Var" columns in Table 3.2):

$$\text{Mean}_{\text{Task}} = \frac{(\text{Pess} + 4\text{ML} + \text{Opt})}{6}$$

$$St.dev_{Task} = \frac{(Pess - Opt)}{6}$$

$$Var_{Task} = St.dev^2 = \frac{(Pess - Opt)^2}{36}$$

The standard deviation of the entire portfolio has to be calculated in a different way. First, all of the project variances have to be added together to obtain the portfolio variance (i.e., 1059.03 in our example). Taking the square root of the portfolio variance yields the project standard deviation (i.e., 32.54) as follows:

$$St.dev_{Portfolio} = \sqrt{Var_{Portfolio}}$$

These two numbers can be utilized by the project manager to establish a link between various effort targets and resulting probabilities of success. The science of statistics tells us that 68.3% of normally distributed population is located within one standard deviation from the population's mean (see Figure 3.2).

Therefore, if the executive team prefers to proceed with two-sided estimates, the resulting output would look like this:

There is a 68.3% chance that the entire project portfolio budget would be between $778,290 and $843,380.

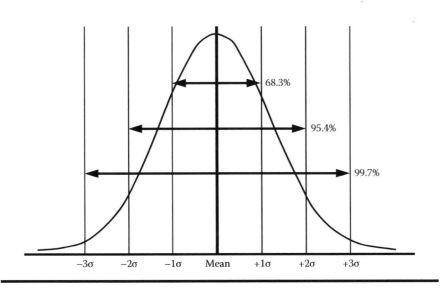

Figure 3.2   Normal distribution—two-sided estimates.

There is a 95.4% chance that the entire project portfolio budget would be between $745,750 and $875,920.

There is a 68.3% chance that the entire project portfolio budget would be between $713,210 and $908,460.

However, managers and customers alike are typically not interested in hearing about the ranges; the typical human being thinks in terms of "is it possible to deliver the project with these resources?"

For those scenarios, Figure 3.3 can be particularly useful. Again, using the mean and standard deviation from our sample portfolio, we can make the following statements:

■ There is a 0.1% chance of successful completion of the portfolio if $713,210 is invested.
■ There is a 0.3% chance of successful completion of the portfolio if $745,750 is invested.
■ There is a 16% chance of successful completion of the portfolio if $778,290 is invested.
■ There is an 84% chance of successful completion of the portfolio if $843,380 is invested.
■ There is a 99.7% chance of successful completion of the portfolio if $875,920 is invested.
■ There is a 99.9% chance of successful completion of the portfolio if $908,460 is invested.

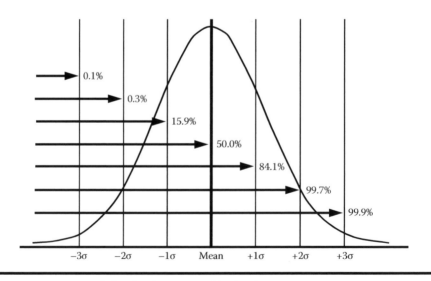

Figure 3.3 Normal distribution—one-sided estimates.

The first two statements are especially interesting. It would be effective to demonstrate using science and proven statistical laws that the probability of finishing this portfolio with $745,750 invested is 0.3%.

Other Things to Consider

How to Improve Your Estimates?

One of the most frequently asked questions by both senior executives and junior project team members alike is, "How can we improve the accuracy of our estimates?"
    The response to this question is based on the three pillars:

1. Detailed understanding of the scope of the work with all the relevant constraints and priorities
2. Access to high-quality, reliable historical data
3. Active involvement of your technical team, or at least your project managers, in preparing estimates

Access to reliable historical data is also an issue because very few companies actually capture key project information. In the author's personal experience, I have heard many excuses from senior managers of various companies why they do not want to capture historical performance data. Some mention lack of understanding of the financial feasibility of such investments. Others claim lack of understanding of the benefits historical data can "bring to the table." Yet another group of a fairly significant size of executives mentions the political issues that could arise from comparing imposed targets and actual results.
    Furthermore, while project managers are encouraged to be proactive and start gathering historical information on their own projects, it usually takes several years of working at the same company and fairly similar projects before one project manager can accumulate a historical database of sufficient size to make any informed and reliable decisions.
    In general (at least), the following historical data should be collected by organizations and project managers alike:

- Total budget
- Total schedule
- Total effort
- Team size
- Scope product delivered
    - Features
    - Requirements
    - Design components
- Type of project

Table 3.3   Common Estimation Oversights

Typical Omissions	Percentage of Total Project Effort
Did we include sick and vacation days?	10–12
Did we include project management tasks?	5–15
Did we include project meetings?	5–15
Did we include documentation tasks?	5–10
Did we include testing tasks?	20–30
Did we include requirements elicitation tasks?	10–15

These methods and their drawbacks lead us to the final estimation improvement technique—a combination of the Wideband Delphi estimation and the PERT methodology.

Common Estimation Oversights

What are some common tasks that are frequently overlooked by project managers, business analysts, and other technical team members? Here is a list of questions to ask the team to find the "popular" omissions during the estimation exercises and the approximate guidelines of the percentages of the total project effort that should be allocated to them (see Table 3.3).

Sample Scenario Analysis

Let us consider an example of how the PERT methodology was used at a European mobile telecommunications company. The organization has established that it would be able to invest approximately 2000 person-months of total effort into its company projects in the upcoming year.

Step 1
After compiling a list of their proposed projects, the executive committee scored them all according to its portfolio scoring model (see Table 3.4).

Note: Projects A and B have been mandated by the local Ministry of Communications and the company head office, respectively, thus making them the mandatory, "joker"-type projects. Hence, both of them received a score of 101 points out of a possible 100.

Table 3.4   Sample Portfolio PERT Estimation—Step 1

Project Name	Project Score
A—Mobile number portability	101[a]
B—Company rebranding	101[a]
C—New tariff plan for students	95
D—Business continuity management	90
E—Talk on the airplane	89
F—Rural network upgrade—District 14	75
G—Rural network upgrade—District 23	74
H—New data plan for professionals	67
I—Mobile financial services	65
J—M2M	62

[a]   "Joker" project that received maximum possible score +1 to take it to the top of the list.

Step 2

The next step involved—with the help of the organization's project managers already familiar with the projects—an assessment of the effort that would be required for each one of the projects (see Table 3.5).

Since adding projects I and J to the project mix would have taken the size of the portfolio more than the 2000 person-month threshold, these initiatives were removed from consideration. However, since the calculations in the table are based on the project effort means, this portfolio had only 50% of projects being delivered with an effort of 1967.5 person-months (see Figure 3.4).

Step 3

The company managers indicated that they are not comfortable with a 50% chance of delivering the projects and requested an increase in the probability of completion. On reviewing the full normal distribution chart for all one-sided estimates (see Figure 3.3), they agreed to target the 84.1% probability of finishing all of the projects. This meant it was necessary to continue removing the projects remaining at the bottom of the list until

$$\text{Mean}_{\text{PERT}} + 1\,\text{St.dev}_{\text{PERT}} \leq 2000 \text{ person-months}$$

Table 3.5 Sample Portfolio PERT Estimation—Step 2

Project Name	Project Score	Opt	ML	Pess	PERT Mean	Cum PERT Mean
A—Mobile number portability	101	300	500	750	**508.33**	508.33
B—Company rebranding	101	100	130	220	**140.00**	648.33
C—New tariff plan for students	95	10	15	25	**15.83**	664.17
D—Business continuity management	90	850	1100	1400	**1108.33**	1772.50
E—Talk on the airplane	89	30	60	75	**57.50**	1830.00
F—Rural network upgrade—District 14	75	50	60	65	**59.17**	1889.17
G—Rural network upgrade—District 23	74	50	60	65	**59.17**	1948.33
H—New data plan for professionals	67	10	20	25	**19.17**	1967.50
~~I—Mobile Financial Services~~	~~65~~	~~40~~	~~70~~	~~90~~	~~**68.33**~~	~~2035.83~~
~~J—M2M~~	~~62~~	~~55~~	~~75~~	~~100~~	~~**75.83**~~	~~2111.67~~

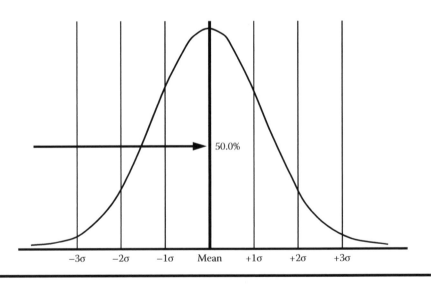

Figure 3.4 Normal distribution—one-sided estimates—50%.

Table 3.6   Sample Portfolio PERT Estimation—Step 3

Project Name	Project Score	Opt	ML	Pess	PERT Mean	PERT Std. Dev.	PERT Var
A—Mobile number portability	101	300	500	750	508.33	75	5625
B—Company rebranding	101	100	130	220	140.00	20	400
C—New tariff plan for students	95	10	15	25	15.83	2.5	6.25
D—Business continuity management	90	850	1100	1400	1108.33	91.7	8402.77
E—Talk on the airplane	89	30	60	75	57.50	7.5	56.25
F—Rural network upgrade—District 14	75	50	60	65	59.17	2.5	6.25
G—Rural network upgrade—District 23	74	50	60	65	59.17	2.5	6.25
H—New data plan for professionals	67	10	20	25	19.17	2.5	6.25
					1967.50	**120.45**	14509.03

After calculating the portfolio's mean and standard deviation using the PERT formulas mentioned earlier in this chapter, the threshold equaled (see Table 3.6)

$$\text{Mean}_{\text{PERT}} + 1 \, \text{St.dev}_{\text{PERT}} = 1967.50 + 120.45 = 2087.95$$

Step 4
Next, "Project H—New data plan for professionals" was removed from the list and the new PERT mean and standard deviation were calculated (see Table 3.7):

$$\text{Mean}_{\text{PERT}} + 1 \, \text{St.dev}_{\text{PERT}} = 1948.33 + 120.43 = 2068.76 \geq 2000$$

This result implied that at least one other project needed to be removed from the list.

Table 3.7 Sample Portfolio PERT Estimation—Step 4

Project Name	Project Score	Opt	ML	Pess	PERT Mean	PERT Std. Dev.	PERT Var
A—Mobile number portability	101	300	500	750	508.33	75	5625
B—Company rebranding	101	100	130	220	140.00	20	400
C—New tariff plan for students	95	10	15	25	15.83	2.5	6.25
D—Business continuity management	90	850	1100	1400	1108.33	91.7	8402.77
E—Talk on the airplane	89	30	60	75	57.50	7.5	56.25
F—Rural network upgrade—District 14	75	50	60	65	59.17	2.5	6.25
G—Rural network upgrade—District 23	74	50	60	65	59.17	2.5	6.25
					1948.33	**120.43**	14502.78

Step 5

It was decided to remove project "Project G—Rural network upgrade—District 23" from the list and recalculate the portfolio's PERT mean and standard deviation (see Table 3.8) as follows:

$$\text{Mean}_{PERT} + 1\ \text{St.dev}_{PERT} = 1889.17 + 120.40 = 2010 \geq 2000$$

However, yet another project had to be removed from the portfolio and, again, the mean and the standard deviation were recalculated.

Step 6

"Project F—Rural network upgrade—District 14" was removed from the list (see Table 3.9).

$$\text{Mean}_{PERT} + 1\ \text{St.dev}_{PERT} = 1830.00 + 120.38 = 1950 \leq 2000$$

This result implied that the desired threshold was reached with a probability of more than 84.1% of finishing all projects with a resource pool of 2000 person-months.

Table 3.8 Sample Portfolio PERT Estimation—Step 5

Project Name	Project Score	Opt	ML	Pess	PERT Mean	PERT Std. Dev.	PERT Var
A—Mobile number portability	101	300	500	750	508.33	75	5625
B—Company rebranding	101	100	130	220	140.00	20	400
C—New tariff plan for students	95	10	15	25	15.83	2.5	6.25
D—Business continuity management	90	850	1100	1400	1108.33	91.7	8402.77
E—Talk on the airplane	89	30	60	75	57.50	7.5	56.25
F—Rural network upgrade— District 14	75	50	60	65	59.17	2.5	6.25
					1889.17	**120.40**	**14496.53**

Table 3.9 Sample Portfolio PERT Estimation—Step 6

Project Name	Project Score	Opt	ML	Pess	PERT Mean	PERT Std. Dev.	PERT Var
A—Mobile number portability	101	300	500	750	508.33	75	5625
B—Company rebranding	101	100	130	220	140.00	20	400
C—New tariff plan for students	95	10	15	25	15.83	2.5	6.25
D—Business continuity management	90	850	1100	1400	1108.33	91.7	8402.77
E—Talk on the airplane	89	30	60	75	57.50	7.5	56.25
					1830.00	**120.38**	**14490.28**

Summary

We started this chapter with a discussion of the challenges faced by executives when conducting portfolio prioritization and resourcing exercises, namely, the inability to come up with precise project-related estimates at the projects' inception.

As a result, we examined two potential approaches to address this problem: the Wideband Delphi and the PERT methodology adjusted for portfolio management rather than for project management needs.

We also discussed the most efficient ways to improve estimation accuracy as well as some common estimation oversights.

Finally, the chapter ended with a step-by-step walk-through of a portfolio resourcing optimization exercise conducted at a European telecom company.

THE APPLICATION: INDUSTRY CASE STUDIES

Chapter 4

Project Portfolio Management in the Pharmaceutical Industry

Pharmaceutical Sector Overview

The pharmaceutical sector represents a fairly unique industry among the multitude of businesses that exist in the world. One of the key factors in the world of pharma is that the development time of an average drug can last for 10–15 years, dwarfing the development times for most technology, software, and even engineering products. Furthermore, pharma research and development (R&D) scientists frequently have to assess and analyze between 5,000 and 10,000 compounds before they are able to synthesize one successful drug (see Figure 4.1).

Another aspect of the pharmaceutical industry is its heavy investment in the new drug R&D. For example, in 2011, the pharma industry spent US$135 billion on R&D. When one compares this amount with other industries, the annual spending by the pharma market is five times greater than that of the aerospace and defense industries, 4.5 times more than that of the chemical industry, and 2.5 times more than that of the software and computer services industry (Joint Research Centre, 2011). It is therefore not surprising that 5 out of 10 global R&D companies are pharma. Also, in 2011, 35 new drugs were launched, while another 3200 are still in development.

Another interesting aspect of the pharmaceutical industry is that it continues to be "socially profitable," so to speak. It is estimated that for every $24 spent on drug development, $89 is saved in healthcare costs worldwide (Ernst and Young, 2012).

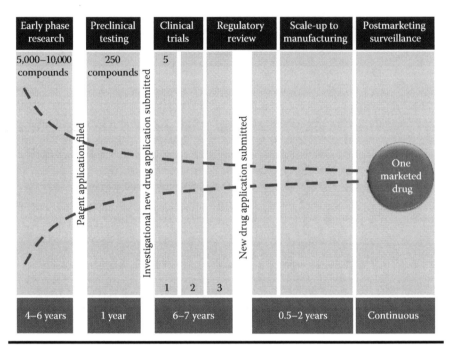

Figure 4.1 Pharmaceutical industry life cycle.

In 2011, 35 new pharmaceutical drugs have been launched around the world. The numbers of drugs in various stages of development around the world are

- Cancer—948
- Cardiovascular—252
- Diabetes—212
- HIV/AIDS—88
- Rare diseases—460

The experts forecast that the pharmaceutical market will grow to US$1200 billion worldwide in 2016. Along with the growth, an interesting shift will occur with respect to the weight of regional powers that constitute the overall pharma market. For example, the so-called emerging markets that typically include Russia, China, Brazil, India, the Philippines, and South Africa used to consume approximately 12% of the global pharmaceutical output in 2005. However, by 2015, they are expected to represent 28% of the global healthcare spending (more than a twofold increase in share).

On the other hand, the share of the United States is expected to fall from 41% of the global market in 2005 to 31% in 2015. By the same token, the European share of the global pharmaceutical consumption will drop from 27% in 2005 to 19% in 2015.

Moreover, many health authorities around the world are now requiring pharmaceutical companies to track and report patients' experiences (referred to as "pharmacovigilance"). These reporting requirements are becoming stricter, raising the investment cost of a given medicine as long as it is being marketed. As a result of this new regulatory requirement, there is an upward pressure applied on R&D costs and a downward pressure on the number of drugs in development:

- 196 drugs developed in 1997–2001
- 146 drugs developed in 2007–2011

One must also consider factors such as the aging world population (especially in developed countries), increased life expectancy, and unhealthy lifestyles that have contributed to the relative growth of noncommunicable diseases such as heart disease, cancer, chronic respiratory diseases, and diabetes.

So, what are the conclusions and lessons learned from these facts and figures? First, the composition of the world pharmaceutical market is shifting from the developed countries (e.g., United States, Canada, and western European countries) toward the developing countries with a large and relatively wealthy population (e.g., Russia, China, India, and Brazil). This implies at least two things: developing new drugs that fit the characteristics of the new markets and establishing closer relationships with local regulatory bodies.

Stringent regulation in the old and new markets, increased competition and the resulting downward pressure on the revenues, and the numbers of new drugs approved per year will force the pharmaceutical industry to become more agile and innovative.

The price for such innovativeness will probably manifest itself in terms of increased risk for new product development projects. As a result, managing these risks properly will present the pharmaceutical companies with significant challenges. Furthermore, some experts argue that pharma companies with well-balanced project portfolios will outperform those that attempt to build their portfolios under the slogan "we will take only winner products" (Ernst and Young, 2012).

Finally, faced with decreasing or stagnant revenues, pharma companies will also try to reduce their development costs probably by focusing on their core competencies rather than venturing into completely new fields.

Pharmaceutical Sector Case Studies

Introduction

In this section, we will examine several different pharmaceutical companies: the first one is a large and financially successful organization, the second one is also a large company that experienced a somewhat stagnant growth in its financials, and the third one is a smaller firm that has to compete with other pharmaceutical giants.

We will also analyze the portfolio models developed by each of these companies and how they fit their strategies as well as their internal and external environments.

European Pharmaceutical Company A

The European pharmaceutical company has a diverse range of products and services. The organization is involved in medical equipment manufacturing, drug development, and hospital management. This company had a strong presence in the European Union (EU) and North American markets.

It had a fruitful decade with its revenues almost tripling and profits growing sixfold in the past 10 years despite the difficult economic times endured by virtually all sectors since the 2008 financial crisis.

This case study represents the results of a working session with the R&D team executives, that is, it does not include "maintenance" or "cost of doing business" projects.

Strategy

The company had a clear global strategy developed by the executive team for the next five years of operation. It included the following goals:

■ Market position expansion—This implied producing more new products for the existing company markets.
■ Extension of the global presence—This implied penetrating new markets with a mix of the existing and new products and services.
■ Improving the innovation—This involved manufacturing cost effectively, leveraging competence in R&D, maintaining a high level of safety, and promoting user-friendliness.
■ Enhance profitability—This mainly involved containing costs.

The Scoring Model

The executive team identified the following key scoring criteria for their R&D projects (see also Table 4.1):

■ Strategic fit
■ Market attractiveness
■ Competitive advantage
■ Technical feasibility
■ Financial (sales)
■ Risk
■ Sales force readiness

Table 4.1   European Pharmaceutical Company A Portfolio Scoring Matrix

Selection Criteria	Points Awarded (Maximum Possible 70)			
	71 Points			
Joker	1 Point	5 Points	10 Points	Kill?
Strategic fit	Fits one of the criteria	Fits two of the criteria	Fits three or four of the criteria	Yes (if score is zero)
Market attractiveness	Number of patients < X	X < number of patients < Y	Number of patients > Y	Yes
Competitive advantage	More than four competitors with similar products	Three or four competitors with similar products	One or two competitors with similar products	Yes (if more than five competitors)
Technical feasibility	Practically no in-house knowledge	Some in-house knowledge	All of the knowledge is in-house	No
Financial (sales)	Revenue < €M	€M < revenue < €N	Revenue > €N	Yes (unless social responsibility)
Risk	High	Medium	Low	Yes (if risk is very high)
Sales force readiness	No knowledge about the proposed product	Some knowledge about the proposed product	Full knowledge about the proposed product	No

The first criterion was the strategic fit since the management team was convinced that all of its research projects must be strongly aligned with the overall company strategy. To introduce measurability to the system, it was decided that a project proposal that fits between three and four of the strategic company goals would get a score of 10 points, whereas proposals reflecting two strategy goals would get a score of 5 points and the ones fitting only one strategic goal, 1 point. This criterion was designated as a "kill" variable. In other words, if a proposed R&D project did not promise to create a new product for the existing market, or introduce a product into a new market, or was not innovative enough, or did not cut costs, it would be automatically removed from the project list without even considering its other aspects.

Market attractiveness criterion was also measurable in a sense that if the marketing team could promise that the potential market would exceed Y number of patients, the project would get a rating of 10 points. If the number of the prospective customers (patients) was anywhere between X and Y, the score received would be 5. And if the prospective patient pool was deemed to be less than X patients, the proposal would get a score of 1. This has also been deemed a "kill" category if the future product promised to address the needs of only a few patients.

Competitive advantage parameters were also designed to be aggressive; a product or service being developed by four or more competitors received a score of 1. Three to four competing companies developing a similar product meant a score of 5 points, whereas the proposed project would get a rating of 10 only if no more than two competitors worked on a similar project. This variable was also included into the "kill" category, if it was known that more than five competing organizations were working on a similar project.

The technical feasibility category is firmly tied to two of the company's strategic goals: innovation and cost control. By awarding the project proposal "bonus" points for requiring in-house knowledge, only the executives hoped to address that issue.

The points for this category were awarded in the following manner:

- Practically no in-house knowledge—1 point
- Some in-house knowledge—5 points
- All of the knowledge is in-house—10 points

The financial (revenue) criterion assessment was fairly straightforward: if the product promised to deliver less than M euros in revenue, it would get a rating of 1 point; if it was deemed to be somewhere between M and N euros, it would get 5 points; and, finally, if the projected revenue was expected to be higher than Y euros, the project would get all 10 points in that category. This variable was also designated as a "kill" category for the cases where it was expected to generate a minimal cash flow. The only exception to this rule was if the company was undertaking this endeavor for "social responsibility" reasons.

The definition of risk for the company executives implied a combination of factors such as the probability of technical success, probability of commercialization, and the probability of project going over budget or being late. After a prolonged attempt to make this category measurable, the executives agreed to go with a "gut feel" for this variable. In other words, the ratings were distributed in the following manner:

- High—1 point
- Medium—5 points
- Low—10 points

The executives also decided that a project proposal would be automatically killed if the overall risk was expected to be exceptionally high. Once more, rather than determining how exactly the "exceptionally high risk" would be estimated, the senior management team decided to go with a "gut feel."

Finally, the senior managers elected to choose a variable fairly rarely seen in the product development companies' portfolio models and called this variable "sales force readiness." What was implied is how ready and knowledgeable its sales people would go out into the market and sell this new product or service to clients. The ranking scores were distributed as follows:

- No knowledge about the proposed product—1 point
- Some knowledge about the proposed product—5 points
- Full knowledge about the proposed product—10 points

The executive committee that participated in the development of this portfolio model decided to employ a "joker" model where a group of very senior managers can approve a project proposal that scored low, but they feel that it may have a serious positive impact on the company's future. Thus, the maximum possible score to obtain in this model is 70 points and the lowest score is 7 points. Designating the project with a "joker" status gives it 71 points to take it to the very top of the portfolio.

Portfolio Balance

The executive committee opted to implement bubble charts to monitor the distribution of the projects in their portfolio (see Figures 4.2 and 4.3):

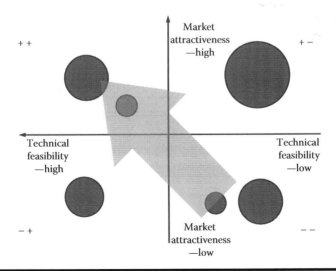

Figure 4.2 European Pharmaceutical Company A portfolio balance—market attractiveness vs. technical feasibility.

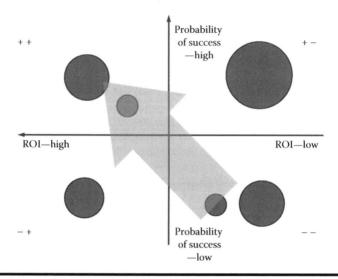

Figure 4.3 European Pharmaceutical Company A portfolio balance—ROI vs. probability of success.

- Market attractiveness vs. technical feasibility
- Return on investment (ROI) vs. probability of success

In addition, they used created pie charts of products and services both by number of specific products in the product group and by their total size in euros.

Strategic Alignment

The executives indicated that in the past they employed an informal model where most of the project proposals were suggested at the "bottom" of their company (i.e., at the department level). This approach prevented the organization from generating breakthrough strategic projects that may have greatly contributed to the bottom line. Thus, they wanted to shift to the top-down, bottom-up model, where the executives would also participate in the R&D project initiation process.

The strategic buckets model was designated to be built in the following manner:

- Breakthrough strategic projects (e.g., new product lines)—20%
- New products—50%
- Improvements to the existing products—30%

European Pharmaceutical Company B

The next company is one of the largest international players in the world market. The organization has two major divisions: pharmaceuticals and diagnostics.

Company assets were in dozens of billions of dollars in 2012, whereas its income was measured in billions of dollars. Despite an overall strong position of the organization, the executives of the company were somewhat concerned with the slow growth in revenues (4%–8% per year) and the net income (1%–3% per year). Consequently, they felt that the company was falling behind the competition and, in the long term, was in danger of losing the leading position in the pharmaceutical industry.

This case study focuses on the organizational R&D projects—both pharmaceutical and diagnostics—and does not include maintenance and staying-in-business ventures.

Strategy

Similar to the previous example, the company executives developed a clear unequivocal strategy that did not involve ambiguous goals. The strategy consisted of four pillars:

1. No "over-the-counter" products—The company decided to avoid the generic drug market altogether and focus on prescription drugs only because of IP protection and higher profit margins.
2. Five research areas—The company decided to focus its R&D efforts on five key pharma fields, including cardiology, cancer, infectious diseases, diabetes, and neuroscience.
3. Focus on personal healthcare—Attending to the physical needs of people who are disabled or otherwise unable to take care of themselves.
4. Personalized drugs—Drugs that can be customized exactly to the needs of a particular patient, including the exact dosage and combination with other medications.

The Scoring Model

The portfolio committee decided to employ the following variables in preparing their scoring model (see Table 4.2):

- Market attractiveness (the number of patients)
- Strategic fit
- Innovativeness
- Risk (both technical and market)
- Effectiveness
- Cannibalization

Table 4.2   European Pharmaceutical Company B Portfolio Scoring Matrix

Selection Criteria	Points Awarded (Maximum Possible 135)			
	136 Points			
Joker	1 Point	5 Points	15 Points	Kill?
Market attractiveness (how many patients are out there?)	Number of patients < X	X < number of patients < Y	Number of patients > Y	Yes
Strategic fit	Fits only one of the strategic fit criteria	Fits two of the strategic fit criteria	Three or four of the strategic fit criteria	Yes (if scores zero)
Innovativeness	Generic approach	Mixed approach	Unique approach	No
Risk (both technical and market)	10% < probability of success < 25%	25% < probability of success < 75%	Probability of success > 75%	Yes (if less than 10%)
Effectiveness	Low	Medium	High	No
Cannibalization	Will compete with several other company drugs	Will compete with 1 other company drug	No competition with other company drugs	No
Core competencies	No in-house knowledge	Some in-house knowledge	All knowledge is in-house	No
Competitors	More than three competitors with similar products	One or two competitors with similar products	No competitors with similar products	No
Financial (revenue)	Revenue < $A	$A < revenue < $B	Revenue > $B	Yes (if revenue minimal)

- Core competencies
- Competitors
- Financial (revenue)

The first category considered was tied directly to the potential number of patients in the market. The project proposal received 1 point for less than X potential patients, 5 points for between X and Y patients, and 15 points for more than Y patients (note the decision to award 15 rather than 10 points for best performance; this way a company can skew its portfolio scoring to reward excellent projects). Furthermore, this category was designated to be a "kill" variable for the project proposals targeting a small number of patients.

In the strategic fit category, the managers, encouraged by the facilitator, decided to use a simple measurable model: 1 point was awarded to proposals that fit only one of the strategic criteria, 5 points to endeavors including two criteria, and 15 points to projects that incorporated three or more of the strategy goals. This criterion also was selected as a "kill" category for projects that did not include any of the strategic goals.

An innovativeness category, a fairly unique variable, was also introduced by the company's managers to focus on the strategic goal of developing more personalized drugs, since this particular field required the organization to partially part ways with the traditional approaches to new drug development. The breakdown of points was as follows:

- Generic approach to drug development—1 point
- Mixed approach to drug development—5 points
- Unique approach to drug development—15 points

The risk factor was included in the model, and the executives decided to incorporate both technical and commercialization aspects into this variable. While this factor will almost always remain a subjective measure, the managers decided to award 1 point for the probability of overall success between 10% and 25%, 5 points for the probability of success between 25% and 75%, and 15 points for the probability of overall success over 75%. Projects with a probability of overall success less than 10% would be killed.

Perceived effectiveness of the drug was another fairly subjective category that was difficult to accurately assess at the beginning of the project. However, it was hoped that as the product development nears its end, it would become more apparent to the managers as to whether the drug possesses the desired effectiveness.

The cannibalization category was introduced to measure the effect of the proposed product on other drugs produced by the company. If the product was expected to compete with more than one of the existing company medicines, it would get a rating of 1 point, if only one medicine 5 points, and products that had no cannibalization effect on any of the company's drugs received 15 points.

The core competencies factor was included with the following parameters:

■ No in-house knowledge—1 point
■ Some in-house knowledge—5 points
■ All knowledge is in-house—15 points

The number of competitors category was included in the model to assess the competitive advantage of the proposed endeavor. The points were distributed as follows:

■ More than three competitors with similar products—1 point
■ One to two competitors with similar products—5 points
■ No competitors with similar products—15 points

And finally, the project revenues category was broken down as follows:

■ Revenue < $A—1 point
■ $A < revenue < $B—5 points
■ Revenue > $B—15 points

Revenue was also designated as a kill category for projects that promised to generate minimal cash flows.

To sum up this model, the maximum number of points a project can generate is 135 and the minimum is 7. Four out of the seven categories have been designated as "kill" factors, making this scoring model an aggressive filtration mechanism.

Furthermore, considering the fairly large number of variables in the model (seven), future assessment and ranking exercises could have become a bit tedious, especially if there was a multitude of project proposals to analyze. Having said that, the executives recognized this fact, but decided to keep all the variables, hoping to calibrate and simplify the model if necessary in the future.

Portfolio Balance

The executive committee decided to track the balance of their portfolio using the following bubble charts (see Figures 4.4 and 4.5):

■ Probability of success vs. total cost
■ Probability of success vs. remaining cost

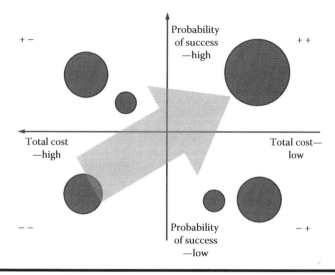

Figure 4.4   European Pharmaceutical Company B portfolio balance—probability of success vs. total cost.

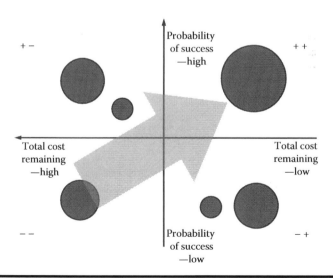

Figure 4.5   European Pharmaceutical Company B portfolio balance—probability of success vs. remaining cost.

Strategic Alignment

The senior management team decided to use the popular top-down, bottom-up model with strategic buckets distributed in the following manner:

- Improvements to existing products—10%
- New products within the existing product families—50%
- New product lines—40%

European Pharmaceutical Company C

This company is somewhat different from those mentioned in the previous two examples. It is a smaller private organization with a strong focus on R&D. It is estimated that between 20% and 40% of the revenue generated by the organization is reinvested into R&D of new drugs.

For example, currently, the company has between 40 and 60 new product projects under way, something that helps the organization to have a presence in many countries in the world and generate billions of dollars in revenues.

Strategy

This company has a unique and an incredibly clear strategy in place. It includes the following measurable objectives:

- The company must achieve at least one market authorization per year.
- At least five of the current company projects should be in Phase 1—development/ therapeutic phase.
- By 2018, 20% of the company revenues should be from Brazil, China, and Russia.

The Scoring Model

The list of scoring variables chosen by the executives of this company was surprisingly short; it consisted of only three criteria (see Table 4.3):

1. Innovativeness (financial benefits vs. risks) based on a comparison of low/ high risks vs. high/low benefits
2. Candidate for China, Brazil, or Russia
3. Resources required to finish the project (measured in person-years)

The first criterion was based on the comparison of the benefits (financial) and risks (technical and commercial) of the proposed project. If the project was expected to generate a relatively low ROI and was associated with high technical and

Table 4.3 European Pharmaceutical Company C Portfolio Scoring Matrix

Selection Criteria	Points Awarded (Maximum Possible 30)			
	31 Points			
Joker	1 Point	5 Points	10 Points	Kill?
Innovativeness (financial benefits vs. risks)	Low benefits and high risks	High benefits or low risks	High benefits and low risks	Yes
Candidate for C/B/R?	Only one of the countries	Any two of the countries	All three of the countries	No
Resources	More than 70 man-years	50–70 man-years	Less than 50 man-years	No

commercialization risks, it would get a score of 1 point. If the proposed endeavor promised a combination of either high risks and high benefits or low risks and low benefits, it would get a score of 5 points. Finally, projects expected to generate a healthy ROI combined with low implementation and commercialization risks would get a score of 10 points. This last category was designated as a "kill" category for project candidates with low ROIs and high risk factors.

The second category was designed to measure the fit of the new drug to the Chinese, Brazilian, or Russian markets. If managers felt that it would be applicable to just one of the markets, the proposal would get a score of 1; if for two markets, 5 points; and for all three markets, 10 points.

Finally, the senior managers felt that a project requiring a total effort of more than 70 person-years should get 1 point in their scoring system. A project estimated to consume between 50 and 70 person-years would get 5 points, whereas an endeavor requiring less than 50 person-years would receive 10 points.

Therefore, the maximum points a project proposal can obtain under this system is 30 and the minimum is 3 points. As in many previous cases, the executives decided to introduce the concept of the "joker" project, where an endeavor that scored low in the matrix, but is being considered the next breakthrough project, would get an automatic score of 31 points, thus taking it to the top of the rank-ordered project list.

Next, two project candidates were scored using the newly developed model.

- Project 1—A proposal to develop Drug A that promised a high ROI and fairly low risks. The product could be sold in China and Russia and required 30–40 person-years of investment.
- Project 2—A proposal to develop Drug B that was a high-benefit, but also a high-risk, endeavor. It could be marketed in Brazil and required 55–70 person-years in resources.

Table 4.4 European Pharmaceutical Company C Project Proposals Comparison

	Drug A	Drug B
Innovativeness (benefits vs. risks)	10 points High benefits and low risks	5 points High benefits
Candidate for C/B/R?	5 points China and Russia	1 point Brazil only
Resources	10 points 30–40 man-years	5 points 55–70 man-years
	25 points	**11 points**

The first project (Drug A) received (see Table 4.4)

- Risk vs. benefits category—10 points
- Geography category—5 points
- Resources category—10 points

The total score for the proposal was therefore 25 points.

On the other hand, second project (Drug B) scored in the following manner:

- Risk vs. benefits category—5 points
- Geography category—1 points
- Resources category—5 points

The total score for the proposal was 11 points.

Portfolio Balance

For the project portfolio balance model, the executives chose to use the traditional benefits vs. risks model (see Figure 4.6).

Strategic Alignment

The executive team chose to proceed with the classical top-down, bottom-up model for the strategic alignment of the portfolio. The strategic buckets were distributed in the following manner:

- Maintenance or stay-in-business projects—20%
- Customization of existing products—30%
- New products and product families—50%

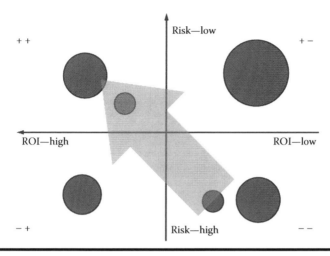

Figure 4.6 European Pharmaceutical Company C portfolio balance—ROI vs. overall risk.

Summary

At the beginning of the chapter, we learned about the following trends in the global pharmaceutical industry:

■ Shifting markets (from North America and EU toward Russia, China, Brazil, India, etc.)
■ Stringent regulations
■ Downward pressure on revenues resulting in the need to control costs
■ Need for more innovation

In the case studies presented, the first European pharmaceutical company (Company A) had three out of four of these goals in its strategy:

1. Extension of the global presence—This implied penetrating new markets with a mix of the existing and new products and services.
2. Improving the innovation—This involved manufacturing cost effectively, leveraging competence in R&D, maintaining a high level of safety, and promoting user-friendliness
3. Enhance profitability—This mainly involved containing costs.

The scoring model included factors such as competitive advantage (direct impact on innovation), technical feasibility, financial impact, and risk (direct impact on the bottom line).

The strong market position of the company was reflected in the moderately aggressive strategic buckets breakdown.

The second company (Company B) also included variables such as innovativeness and competitor analysis to focus on innovation and core competencies and revenue to strengthen its financial position.

The aggressive balance of the strategic buckets, with 90% of the resources going to new products and product families, shows the commitment of the organization's executives to break out of the stagnant situation for the previous several years and significantly increase the company's sales and profits.

The situation of the third company (Company C) was quite different from the first two; it was much smaller in size—both revenues and assets—and regardless positioned itself in such a way to be able to compete with larger pharmaceutical firms.

Hence, the directors chose a strategy focused on maximizing the number of drugs approved, emerging markets, and favoring projects that required less rather than more resources. This approach was coupled with a strong skew toward new product development in the portfolio strategic alignment, which allowed them to come up with a fairly unique but potentially highly effective portfolio model.

Chapter 5

Project Portfolio Management in the Product Development Industry

Product Development Sector Overview

The product development industry is one of the most challenging topics to write about and more importantly generalize. It is a vast area covering almost every aspect of our lives. Product development companies are responsible for the creation and production of items such as cars, satellites, drugs, computers, phones, shampoos, and software, just to name a few.

It is also difficult to provide any numbers with respect to the product industry. Based on a simple statistic that the United States alone has exported $2.1 trillion worth of goods in 2011 (CIA World Factbook, 2014), we can only surmise the worldwide numbers for the total goods production being at between $8 trillion and $9 trillion, based on the fact that U.S. economy represents approximately one-quarter of the global.

Therefore, it is fairly difficult to generalize any aspect of the product development sector because of its vastness and diversity. However, various experts agree more or less that there are several challenges that need to be overcome in the upcoming several years (Yakimov and Woolsey, 2010).

First, the economy is becoming more and more globalized with competition, frequently and unexpectedly, "popping up" in different parts of the world.

Second, the emergence of new markets, especially India, China, Russia, and Brazil, now implies that organizations should tailor new products that should be tailored specifically for those regions.

Third, considering the facts mentioned earlier (competition and new markets), innovation is becoming more and more important for the companies to survive. Time to market and competitive pricing are also natural by-products of globalization.

Finally, the protection of intellectual property (IP) in the age when a successful product can be broken apart and reverse-engineered and manufactured in a completely different part of the world is also an important issue.

Interestingly enough, in a survey conducted recently by Planview Software (Appleseed Partners and OpenSky Research, 2013), many of the aforementioned challenges have been echoed by executives in the product development companies. According to this report, between 40% and 75% of those surveyed, innovation growth is imperative, but their companies lack a clear path on how to improve and develop in this area.

Furthermore, more than 50% of respondents in this survey indicated their product portfolios are not as well aligned with their company's strategies and objectives as they should be. Combine this with 53% of respondents who thought that they had more work than people to do it (53%), and 54% were convinced that they are unable to drive innovation fast enough to meet market demands.

So, in the light of the pressures exerted from the external factors and the internal challenges of the companies, what should they do to properly prepare for the future?

To start with, they must innovate constantly to adapt to economic and technological changes. To achieve this goal, their research and development (R&D) spending must be properly reflected in the company strategy.

They should also make an effort to diversify their product portfolios to fit the needs of different markets, including the emerging markets in Brazil, China, India, and Russia.

Furthermore, they must green technologies to decrease costs to make their products more competitive in the global marketplace. This is especially relevant for the organizations operating primarily in the developed countries as their human costs tend to be higher than those in the developing countries.

Product Development Sector Case Studies

Introduction

In this chapter, we will examine the portfolio models created by a bearings manufacturer, a software producer, a rail transport company, a medical equipment manufacturer, a food packaging company, a satellite operator, and a clothing manufacturer.

We will also analyze the portfolio models developed by each of these companies and how they fit their strategies as well as their internal and external environments.

Company A: Bearings Manufacturer

The first company to be discussed in this chapter is a successful bearings manufacturer. For a number of years, it has focused all of its R&D efforts solely on the bearings production.

However, under considerable pressure from its sales team, the executives decided to review their strategy. The sales department has insisted several times that when their staff talk to their customers, they keep asking questions about other bearing-related products such as sealants, lubricants, and electronic components, which the company was not producing.

Strategy

As a result of the aforementioned events, the executives prepared a new strategy for the R&D department that consisted of the following initiatives:

- Develop new (i.e., lubricants, sealants, and electronic components) product families.
- Develop attractive products (i.e., something that customers demand).
- Increase revenues and profitability by developing new product families.
- Increase market share in the new markets.

Scoring Model

The scoring model the executives developed includes the following variables (see Table 5.1):

- Strategic fit
- Possible synergies
- Financial value
  - Payback
- Technical complexity (skills in-house)
- Market attractiveness
- Competition and IP

Not surprisingly, the first variable added to the prioritization matrix was strategic alignment. If the project proposal fit one of the strategic criteria, it would

Table 5.1　Company A: Bearings Manufacturer Portfolio Scoring Matrix

Selection Criteria	Points Awarded			
	91 Points			
Joker	1 Point	5 Points	15 Points	Kill?
Strategic fit	Low Fits one of the criteria	Medium Fits two or three of the criteria	High Fits four or more of the criteria	Yes, unless a "joker" project
Possible synergies	Low Cannot combine sales of the proposed product with other product families	Medium Can combine sales of the proposed product with one other product family	High Can combine sales of the proposed product with two or more other product family	No
Financial value	Minor 0 < NPV < $1 million	Medium $1 million < NPV < $5 million	Major NPV > $5 million	Yes, unless a "joker" project
Technical complexity	Very difficult A significant external expertise is required	Somewhat difficult Will need some external expertise	Easy Can be implemented by internal employees	No
Market attractiveness	Low Few requests	Medium Several requests	Major Many requests	Yes, unless a "joker" project
Competition and IP	High Many competitors Weak IP protection	Medium three or four competitors Normal IP protection	Low zero or two competitors Strong IP protection	No

receive 1 point; if the proposal covered two or three of the strategies, 5 points; and finally, 15 points if it addressed four or more of the strategic criteria. In addition, this variable has been designated as a "kill" category. In other words, if the proposed venture did not address any of the strategic criteria, it automatically became a candidate for dismissal, unless awarded a "joker" status by the senior managers.

The points for the next variable—possible synergies—have been distributed in the following manner:

- Cannot combine sales of proposed product with other product families— 1 point
- Can combine sales of proposed product with 1 other product family— 5 points
- Can combine sales of proposed product with 2+ other product families— 15 points

The executives felt strongly that in order for the company to maintain its competitive position in the market, they needed to add a variable to account for the financial gains to be realized from new products. After much consideration, the executives chose net present value (NPV) as the third variable to be added to the matrix. The project would receive 1 point for NPV less than 1 million, 5 points for NPV between $1 and $5 million, and 15 points for NPV exceeding $5 million. This category was deemed to be a "kill" variable with the proposals where NPV was less than zero, requiring special C-level approval.

The managers also felt that the technical complexity of the proposed ventures should also be considered as the company was planning to move into new technologies, including lubricants, sealants, and electronic components. They added this variable to the model with the following parameters:

- A significant external expertise is required—1 point
- Will need some external expertise—5 points
- Can be implemented by internal employees—15 points

Considering the sales team initiated this transformation at the R&D department, the executives added the market attractiveness variable to the mix to gauge the potential sales of the product. The sales executives proposed to measure this category in terms of customer requests for a specific product type. This decision meant the products that received only a few inquiries would get 1 point; the ones with several requests, 5 points; and the ones receiving a lot of inquiries from the clients, 15 points. In addition, this variable has also been designated as a "kill" category for the products receiving few or no requests at all.

Finally, the managers included the competition and IP category to ensure that the new products are unique and could be protected by patents. The points were distributed as follows:

- Many competitors and weak IP protection—1 point
- Three to four competitors and normal IP protection—5 points
- Zero to two competitors and strong IP protection—15 points

This scoring approach meant a candidate project could generate a maximum of 90 points and a minimum of 6 points—unless it scored zero in one of the "kill" categories.

Portfolio Balance

The executives indicated that they would be interested in analyzing their portfolio from the market or technical risk vs. reward balance perspective (see Figure 5.1).

Strategic Alignment

Considering the situation and the new strategy of the company's R&D department, the managers decided to designate the following resource buckets for the company's upcoming projects:

- 10%—Bearings
- 30%—Sealants
- 30%—Lubricants
- 30%—Electronic components

In addition, they adopted the "top-down, bottom-up" approach for all project proposals.

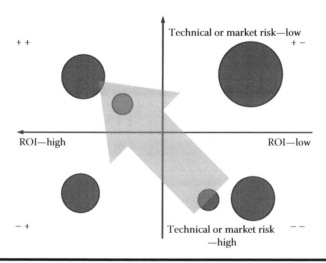

Figure 5.1   Company A: Bearings manufacturer portfolio balance—market or technical risk vs. reward.

Company B: Software Producer

The second company in this chapter is the software developer that produced various applications mainly for the telecommunication industry. While the company has been experiencing an aggressive growth in the past five years, there were several issues identified by the senior management as follows:

- We continue to be a one-product company.
- We focus on just several geographical markets.
- We have been growing too fast people-wise but not fast enough revenue-wise.
- The telecom industry worldwide is saturated with various value-added services software.

Strategy

Because of these challenges, the managers decided on the following strategic initiatives for the next three-year period:

- Expand our product family.
- Expand into new geographic markets, especially Asia and South America.
- Reduce the headcount expansion.
- Enable higher revenue growth.
- Expand into nonmobile industries (e.g., utilities, banking, transportation).

Scoring Model

The scoring model developed during the facilitated session with the executive team contained the following five variables (see also Table 5.2):

1. Strategic fit
2. Leverage of core competencies
3. Financial forecast
4. Market attractiveness
5. Synergy with other projects/products

The points for the first variable added to the model—the strategic fit—were distributed as follows:

- Fits one of the criterion—1 point
- Fits two of the criteria—5 points
- Fits three or more of the criteria—10 points

Table 5.2 Company B: Software Producer Portfolio Scoring Matrix

New Product Projects	Points Awarded			
	51 Points			
Selection Criteria	1 Point	5 Points	10 Points	Kill?
Strategic fit	Low Fits one of the criteria	Medium Fits two of the criteria	High Fits three or more of the criteria	Yes, if the proposed project fits zero of the strategic fit criteria, it is removed from further consideration
Leverage of core competencies	Low Completely new technologies and brand-new domain for the company	Medium The technologies involved and the domain are somewhat familiar to the company	High Both the technologies involved and the domain knowledge are familiar to the company's employees	Yes, if the proposed project scores very low on a core competencies criterion, it is removed from further consideration
Financial forecast[a]	Minor 5% < ROI (IRR) < 10%[b]	Medium 10% < ROI (IRR) < 20%	Major ROI (IRR) > 20%	No
Market attractiveness What is the market size in terms of the number of companies in the world? How many companies can potentially be targeted with this product?	Low Between 25 and 49 companies	Medium Between 50 and 149 companies	High More than 150 companies	Yes, if the number of companies <25

(Continued)

Table 5.2 (*Continued*)   Company B: Software Producer Portfolio Scoring Matrix

New Product Projects	Points Awarded			
Selection Criteria	51 Points			
	1 Point	5 Points	10 Points	Kill?
Synergy with other projects/ products Can this product be cross-sold with other company's offerings? Can the technology know-how be borrowed from other company projects?	Low The product can be cross-sold with one company product and/ or one technology know-how can be borrowed	Medium The product can be cross-sold with two company products and/or two technology know-hows can be borrowed	High The product can be cross-sold with three or more company products and/or three technology know-hows can be borrowed	No

a Also consider the absolute value of the expected revenue.
b An ROI of less than 5% requires a special executive management approval.

This variable was designated as a "kill" category with projects that were not aligned with any of the strategies receiving an automatic score of zero and being removed from the list, unless they fell into the "joker" category.

The points for the "leverage of core competencies" category were assigned as follows:

- Completely new technologies and a new domain for the company—1 point
- The technologies involved and the domain are somewhat familiar to the company—5 points
- Both the technologies involved and the domain knowledge are familiar to the company's employees—10 points

Again, this factor was assigned to the "kill" category for the proposals completely outside of the company's domain knowledge.

The return on investment (ROI) points were awarded in the following manner:

- 5% < ROI (internal rate of return [IRR]) < 10%—1 point
- 10% < ROI (IRR) < 20%—5 points
- ROI (IRR) > 20%—10 points

An interesting aspect of this category is that while it has not been designated as a "kill" variable, the team required special approval from the executives for the projects with an ROI less than 5%.

Market attractiveness was the fourth variable added to the model with the points allocated based on the potential number of companies that can be targeted with the proposed product as follows:

- Between 25 and 49 companies—1 point
- Between 50 and 149 companies—5 points
- More than 150 companies—10 points

If the proposed product could target less than 25 companies worldwide, the proposal would receive an automatic score of zero and was removed from the list.

Finally, the executives decided to include the "synergy with other products" variable to promote cross-selling between the new and the existing company offerings as follows:

- The product can be cross-sold with one company product and/or one technology know-how can be borrowed—1 point
- The product can be cross-sold with two company products and/or two technology know-hows can be borrowed—5 points
- The product can be cross-sold with three or more company products and/or three technology know-hows can be borrowed—10 points

Portfolio Balance

The executive team chose to monitor the portfolio balance using the ROI vs. resources bubble chart (see Figure 5.2).

Strategic Alignment

The company adapted the "top-down, bottom up" approach to the project selection. The strategic buckets distributed the resources in the following manner:

- Maintenance—5%
- Product improvements—20%
- New products—75%

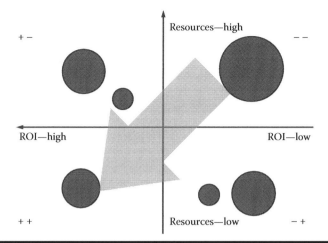

Figure 5.2   Company B: Software producer portfolio balance—ROI vs. resources.

Company C: Rail Transport Company

The next company is a rail transport engineering company that has encountered several challenges in the past several years. The organization has been reporting heavy losses from its operations for the past decade with no sign of potential improvement.

The analysis of the company's operations showed that one of the main reasons for the poor performance of the company was the large number of products produced by the organization, which led to various customization requests from their customers.

The company then had a large number of concurrent projects with a majority of them being customizations rather than new product development ventures. As a result, the quality of the project's products has also declined, leading to major delays in the product delivery to customers.

Strategy

This situation led to the executives determining the following strategy:

- Implement a rigorous project portfolio management system to (a) prioritize projects and (b) cut low-priority ventures.
- Create platform products to decrease the degree of customization and to eliminate complexity.
- Increase sales and margins per product category.
- Expand the markets to China, Africa, and South America.
- Improve customer care.
- Improve the product quality.

Scoring Model

The scoring model that they developed as a result of the project portfolio management initiative consists of the following six variables (see also Table 5.3):

1. Market attractiveness
2. Fit to existing supply chain
3. Product and competitive advantage
4. Technical feasibility
5. Time to break even
6. NPV

Interestingly enough, the executives decided not to include the strategic fit as one of the variables in the model because they felt that the combination of the variables selected would address all of their strategic initiatives in a more efficient way.

The first variable added to the model was the proposed product's market attractiveness. The points were distributed in the following manner:

- Only a few requests from current customers and/or the market is declining—1 point
- A considerable number of requests from current customers and/or the market is stable—5 points
- Numerous requests from current customers and/or the market is growing—10 points

A kill category was any project proposals scoring low (i.e., few requests from customers and declining market), which received an automatic score of zero and was removed from further consideration.

The second variable was added to simplify the addition of new products and product families to the existing portfolio. The fit to the existing supply chain was ranked in the following manner:

- Major changes are required to the existing supply chain—1 point
- Some changes are required to the existing supply chain—5 points
- Very few or no changes are required to the existing supply chain—10 points

Product and competitive advantage was the next factor considered when adding projects to the portfolio. If the product uniqueness and desirability were low, the proposal would receive 1 point. If the venture's product was somewhat unique and desirability was average, it would receive 5 points. Finally, if the product was unique and its desirability was high, it was awarded 10 points.

If the product uniqueness and benefit for customers were low, the project would be awarded a score of zero and would be removed from the portfolio.

Table 5.3   Company C: Rail Transport Company Portfolio Scoring Matrix

Selection Criteria	Points Awarded			
	61 Points			
Joker	1 Point	5 Points	10 Points	Kill?
Market attractiveness	Low Only few requests from current customers Market is declining	Medium Considerable number of requests from current customers Market is stable	High A lot of requests from current customers Market is growing	Yes, if very low and not a regulatory or "joker" project
Fit to the existing supply chain	Poor Major changes are required to the existing supply chain	Medium Some changes are required to the existing supply chain	Excellent Very few or no changes are required to the existing supply chain	No
Product and competitive advantage	Poor Product uniqueness and desirability are low	Medium Provides somewhat unique and desired products to the customers	Excellent Provides highly unique and desired products to the customers	Yes, if very low, unless a regulatory or "joker" project
Technical feasibility	Low Complex project involving a lot of external expertise	Medium Somewhat complex project with certain degree of outsourcing involved	High Relatively simple project and no or little outsourcing is required	Yes if very complex, unless a regulatory or "joker" project
Time to break even	T > 6 years	3 < T < 6 years	T < 3 years	Yes, if T > 10 years, unless a regulatory or "joker" project
NPV	NPV < €4 million	4 < NPV < €20 million	NPV > €20 million	No

As in many previous portfolio models, the next variable added to the mix was the technical feasibility of the proposed project, with highly complex projects receiving the score of 1 point; the medium difficulty ones, 5 points; and the easy ones, 10 points. The projects that were deemed to be highly complex would be awarded a score of zero and would be removed from further consideration.

Time to break even points were distributed in the following fashion:

- T > 6 years—1 point
- 3 < T < 6 years—5 points
- T < 3 years—10 points

Furthermore, projects with projected payback times exceeding 10 years would be automatically removed from the portfolio.

Finally, to boost the company's financial performance, the NPV was added to the portfolio mix as follows:

- NPV < €4 million—1 point
- 4 < NPV < €20 million—5 points
- NPV > €20 million—10 points

With this model, the maximum number of points a project could generate was 60, while the minimum—unless it was added to the kill category—was six.

Portfolio Balance

The company's executives decided to monitor the balance of their portfolio via the risk vs. reward graph (see Figure 5.3).

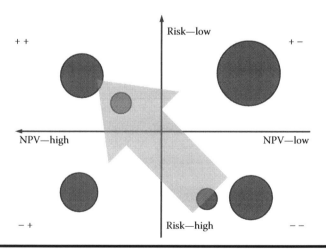

Figure 5.3   Company C: Rail transport company portfolio balance—NPV vs. risk.

Strategic Alignment

The managers decided to adopt the "top-down, bottom-up" approach to the project selection with the following strategic buckets:

- Stay-in-business projects—10%–20% of the total project expenses
- Product improvements—60%–70% of the total project expenses
- New product lines—10%–30% of the total project expenses

Company D: Medical Equipment Manufacturer

The next company in this chapter is that develops and manufactures solutions for the medical industry including blood sampling and blood gas analysis equipment. At the time of the project portfolio management development exercise, the organization was doing well with both revenues and profits steadily growing over the previous several years.

However, the company's managers felt that they had been neglecting new emerging markets and should dedicate more time and resources to the development of new products and product platforms to expand its operations in China, Brazil, and India.

In addition, the project's team complained several times that resources were needed to address both the existing product improvement—the so-called technical debt—projects and the new product development ventures.

Strategy

The executives prepared the following strategies:

- Improve usability and reliability (i.e., eliminate technical debt) of our existing products.
- Develop new features for the existing products targeting developed countries.
- Develop new platforms for the China, India, and Brazil markets.

Scoring Model

They developed a fairly simple scoring model with only the following four variables (see also Table 5.4):

1. Strategic fit
2. Financials (NPV)

Table 5.4 Company D: Medical Equipment Manufacturer Portfolio Scoring Matrix

Selection Criteria	Points Awarded			
	61 Points			
Joker	1 Point	5 Points	15 Points	Kill?
Strategic fit	Low Fits only one of our strategies	Medium Fits two of our strategies	High Fits all three of our strategies	Yes, if does not fit any of the strategies and not a regulatory or "joker" project
Financials (NPV)	Poor NPV < US$1 million	Medium US$1 < NPV < US$9 million	Excellent NPV > US$9 million	No
Market attractiveness	Poor Product uniqueness and desirability are low	Medium Provides somewhat unique and desired products to the customers	Excellent Provides highly unique and desired products to the customers	Yes, if very low, unless a regulatory or "joker" project
Technical feasibility	Low Complex project involving a lot of external expertise	Medium Somewhat complex project with certain degree of outsourcing involved	High Relatively simple project and no or little outsourcing is required	No

3. Market attractiveness
4. Technical feasibility

The executives decided to add the strategic fit as the first variable to the scoring matrix. Each proposal would receive 1 point if it fits one of the corporate strategies, 5 points if it fits two of the strategies, and 15 points if it fits all three of the strategic initiatives. If projects did not fit any strategies, they were removed from further consideration unless they were designated as "jokers" by the executives.

The points for the NPV were distributed in the following manner:

- NPV < US$1 million—1 point
- US$1 < NPV < US$9 million—5 points
- NPV > US$9 million—15 points

Market attractiveness of the proposed product was the next variable in the scoring model with the points distributed as follows:

- Product uniqueness and desirability is low—1 point
- Provides somewhat unique and desired products to the customers—5 points
- Provides highly unique and desired products to the customers—15 points

If a proposed project scored very low in this category, it was removed from further consideration unless it was deemed to be a "joker" project.

Finally, the managers felt that to lower the burden on the product development teams, they should reward simpler, less complex projects and penalize more complicated ones. As a result, they added the technical feasibility variable to the model, where complex projects received 1 point; medium difficulty ones, 5 points; and the simple endeavors, 15 points.

Portfolio Balance

The managers chose to monitor their project portfolio balance using the NPV vs. cost chart to promote smaller, less sophisticated projects (see Figure 5.4).

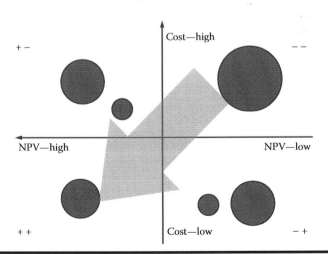

Figure 5.4   Company D: Medical equipment manufacturer portfolio balance— NPV vs. cost.

Strategic Alignment

The managers decided to proceed with the "top-down" projects' selection methodology, where the project proposals would only be generated by the senior managers. The proposed bucket split looked as follows:

- Stay-in-business projects—10%
- Product improvements—70%
- New product lines—30%

Company E: Food Packaging Company

The next organization is a food packaging company that operates in several dozen countries. As of 2014, this company, although successful financially, was facing several challenges. First, the managers felt that it was losing the market share for the canning solutions to its competitors, mainly because this sector has been overlooked from the investment perspective. Second, they realized that rising operational costs would present a potential problem in the long run. And, finally, the executives felt that the markets that it was developing were becoming saturated, and it needs to focus on capturing market shares in several developing countries—A, B, C, and D.

Strategy

The executives came up with a three-pronged strategy for the organization as follows:

1. Grow its canning business by developing new canning products and solutions.
2. Reduce operational costs.
3. Focus on geographical areas A, B, C, and D.

Scoring Model

The executive board's scoring model contained the following variables (see also Table 5.5):

- Strategic fit
- Time to market
- Market attractiveness
- Technical feasibility
- Competitive advantage
- Fit to the existing supply chain

Table 5.5 Company E: Food Packaging Company Portfolio Scoring Matrix

Selection Criteria	Points Awarded			
	91 Points			
Joker	1 Point	5 Points	15 Points	Kill?
Strategic fit	Low Fits only one of our strategies	Medium Fits two of our strategies	High Fits all three of our strategies	Yes, if does not fit any of the strategies and not a regulatory or "joker" project
Time to market	High T > 5 years	Medium 2 < T < 5 years	Low T < 2 years	No
Market attractiveness	Poor Low expected volume sales and/or does not include "must win" customers	Medium Medium expected volume sales and/or may include "must win" customers	Excellent High expected volume sales and/or most likely will include "must win" customers	Yes, if very low, unless a regulatory or "joker" project
Technical feasibility	Low Complex project involving a lot of external expertise	Medium Somewhat complex project with certain degree of outsourcing involved	High Relatively simple project and no or little outsourcing is required	No
Competitive advantage	Low More than four competitors offering similar products Low probability of getting a patent	Medium Between two and three competitors offering similar products Medium probability of getting a patent	High Between zero and one competitors offering similar products High probability of getting a patent	No

(Continued)

Table 5.5 (*Continued*) Company E: Food Packaging Company Portfolio Scoring Matrix

Selection Criteria	Points Awarded			
	91 Points			
Joker	1 Point	5 Points	15 Points	Kill?
Fit to the existing supply chain	Low Major changes to the existing supply chain will be required	Medium Some changes to the existing supply chain will be required	High No or very few changes to the existing supply chain will be required	No

The first variable added to the scoring matrix was the strategic fit. Since the company's strategy consisted of only three initiatives, the points were easily distributed in the following fashion:

1. Fits only one of our strategies—1 point
2. Fits two of our strategies—5 points
3. Fits all three of our strategies—15 points

If project did not fit to any of these strategic initiatives, it received an automatic score of zero and was removed from further consideration unless it was a "joker" or regulatory endeavor.

Time to market was the next variable added to the model since managers wanted to encourage short-term rather than long-term projects. Proposals with an expected duration of more than five years received 1 point; the ones with the durations between two and five years, 5 points; and the ones with a duration of less than two years, 15 points.

Market attractiveness points were distributed as follows:

■ Low expected volume sales and/or does not include "must win" customers—1 point
■ Medium expected volume sales and/or may include "must win" customers—5 points
■ High expected volume sales and/or most likely will include "must win" customers—15 points

If a project had a very low level of market interest in the new product and no apparent "must win" customers, it was removed from further consideration.

The next category that the managers wanted to consider when prioritizing their project proposals was the perceived technical complexity of the project. The points in the model were distributed in the following fashion:

- Complex project involving extensive external expertise—1 point
- Somewhat complex project with a certain degree of outsourcing involved—5 points
- Relatively a simple project with little or no outsourcing required—15 points

Competitive advantage was the fifth variable added to the model to invigorate the innovativeness at the company. The points were distributed as follows:

- More than four competitors offering similar products; low probability of getting a patent—1 point
- Between two and three competitors offering similar products; medium probability of getting a patent—5 points
- Between zero to one competitors offering similar products; high probability of getting a patent—15 points

Finally, the executives included the "fit to the existing supply chain" variable to align the company's new products with its existing distribution channels. The products requiring major changes to the supply chain would receive 1 point; the ones that required some changes, 5 points, and the ones requiring limited or no changes to the supply chain, 15 points.

Portfolio Balance

The senior managers decided to monitor the balance of the portfolio by using the "risk vs. reward" bubble chart (see Figure 5.5).

Strategic Alignment

The managers selected the "top-down, bottom-up" approach for the portfolio alignment and designated the following resource buckets for their projects:

- Regulatory, maintenance, and "stay-in-business" projects—20%
- Product improvements—40%
- New products—40%

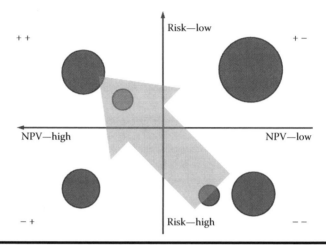

Figure 5.5   Company E: Food packaging company portfolio balance—NPV vs. risk.

Company F: Satellite Operator

The sixth company is a satellite operator and producer that operates several dozen satellites, providing communication services to businesses and government agencies and broadcasts TV and radio channels to audiences worldwide.

Despite an acceptable financial performance, the company executives felt that there were several challenges awaiting the organization in the near future. One of the potential problems was saturated existing markets and stiff competition from other satellite operators. As a result, the company could not assure the same growth rates as that was achieved in the previous years.

The organization needed to move into the new emerging markets and develop new products and services for both developed and developing countries.

Strategy

Because of these challenges, the executives created the following strategic plan for the upcoming five years:

- Maintain market shares in the developed countries.
- Increase investments in the developing countries.
- Improve innovation.
- Improve vertical integration of the organization (develop end-to-end solutions).

Scoring Model

The executives developed the scoring model with the following six variables (see also Table 5.6):

1. Strategic alignment
2. Customer need
3. Synergies with the existing business
4. Technical feasibility
5. Profitability (payback)
6. Commercial/technical risk

The first variable added to the model was the popular "strategic alignment" category. The points were allocated in the following fashion:

- Fits one criterion—1 point
- Fits two to three criteria—5 points
- Fits four criteria—10 points

If a project reflected none of the company strategies, it was automatically removed from the list, unless it was designated as a "joker" or a regulatory project.

Customer need was the next category added to the scoring matrix. If there were only a few requests for the product from the sales department, the project would receive a score of 1 point; if there were several requests, 5 points; and, finally, products requested by many customers, 10 points.

The third variable included was "synergies with the existing business" to align the future product with the existing company offerings. The points were distributed in the following fashion:

- Difficult to cross-sell the new product with the existing product lines—1 point
- Fairly easy to cross-sell the new product with the existing product lines—5 points
- Very easy to cross-sell the new product with the existing product lines—10 points

Technical feasibility was the next variable added to the model to steer the company away from larger, complicated projects. The points in this category were awarded as follows:

- Complex project involving extensive external expertise—1 point
- Somewhat complex project with a certain degree of outsourcing involved—5 points
- Relatively simple project with little or no outsourcing required—10 points

Table 5.6   Company F: Satellite Operator Portfolio Scoring Matrix

Selection Criteria	Points Awarded			
	51 Points			
Joker	1 Point	5 Points	10 Points	Kill?
Strategic alignment	Low Fits one of the criteria	Medium Fits two or three of the criteria	High Fits four of the criteria	Yes, unless a regulatory or "joker" project
Customer need	Low Few customer requests coming from the sales department	Medium Certain number of customer requests coming from the sales department	High Certain number of customer requests coming from the sales department	No
Synergies with the existing business	Low Difficult to cross-sell the new product with the existing product lines	Medium Fairly easy to cross-sell the new product with the existing product lines	High Very easy to cross-sell the new product with the existing product lines	No
Technical feasibility	Low Complex project involving a lot of external expertise	Medium Somewhat complex project with certain degree of outsourcing involved	High Relatively simple project and no or little outsourcing is required	Yes if very complex, unless a regulatory or "joker" project
Profitability (payback)	Low T > 5 years	Medium 1 < T < 5 years	High T < 1 year	Yes, if T > 10 years, unless a regulatory or "joker" project

The projects that were complicated with an extensive outsourcing involved were removed from further consideration, again, unless they were designated as "joker" or regulatory endeavors.

Finally, to promote financially attractive ventures, the executives included the payback variable to the model with the points distributed in the following manner:

- $T > 5$ years—1 point
- $1 < T < 5$ years—5 points
- $T < 1$ year—10 points

Any projects with a payback of more than 10 years were removed from further consideration unless they were "joker" or regulatory endeavors.

Portfolio Balance

The executive team decided to monitor the portfolio performance using the risk vs. reward diagram (see Figure 5.6).

Strategic Alignment

The management team decided to use the popular top-down, bottom-up model with the following designated resource buckets:

- Mandatory and maintenance projects—20%
- Product improvement projects—40%
- New product line projects—40%

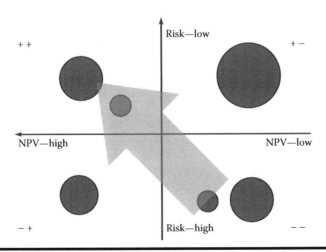

Figure 5.6   Company F: Satellite operator portfolio balance—NPV vs. risk.

Company G: Clothing Manufacturer

The final company is a multinational clothing manufacturer operating worldwide with close to 2000 stores. While the financial performance of the company was fairly respectable, the executives could foresee several potential challenges in the long run.

First, they felt that the company was concentrating too much on the retail side of the business and limiting the investments in the online and mobile stores. Also, for more than several years, the expenses of the organization grew at a higher rate than its revenues (this does not track with the aforementioned statement about the respectable financial performance). Finally, the managers felt that the company was ignoring the emerging markets and should spend more time in developing products for markets in Brazil, China, Russia, and India.

Strategy

As a result of the challenges outlined earlier, the executives developed a four-pronged strategy as follows:

1. Continue the development of the multichannel business with more attention dedicated to online stores, mobile stores, and mobile app stores.
2. Remain brand led and consumer focused.
3. Remain a high-performance organization (increase revenue and decrease costs).
4. Focus on Brazil, China, Russia, and India.

Scoring Model

The scoring matrix developed during a facilitated project portfolio management workshop contained only the following four variables (see also Table 5.7):

1. Strategic alignment
2. Financial benefit (NPV)
3. Resource needs
4. Risk if not executed

The strategic alignment variable was added to the model to make sure that the projects supporting company strategic initiatives received higher standing.

The financial benefit (NPV) and resources needed were included in the model to promote smaller ventures that would improve the company's financial performance.

Finally, the executive committee decided to include the "risk if not executed" variable to the model to promote the projects that were critical to the existence of the company.

Table 5.7   Company G: Clothing Manufacturer Portfolio Scoring Matrix

Selection Criteria	Points Awarded			
	41 Points			
Joker	1 Point	5 Points	10 Points	Kill?
Strategic alignment	Low Fits one of the criteria	Medium Fits two of the criteria	High Fits three or more of the criteria	Yes, unless a regulatory or "joker" project
Financial benefit (NPV)	Low NPV < US$20 million	Medium 20 < NPV < US$50 million	High NPV > US$50 million	No
Resource needs	High R > 1000 man-days	Medium 300 < R < 1000 man-days	Low R < 300 man-days	No
Risk if not executed	Low Failure to implement the project carries little or no risk (e.g., operational, reputational, regulatory)	Medium Failure to implement the project carries some risk (e.g., operational, reputational, regulatory)	Low Failure to implement the project carries major risk (e.g., operational, reputational, regulatory)	No

Portfolio Balance

The executives decided to monitor the balance of their portfolio by employing the risk vs. reward bubble chart (see Figure 5.7).

Strategic Alignment

The managers decided to use the "top-down, bottom-up" model with the following strategic buckets designated for the company's project portfolio:

■ Maintenance and regulatory projects—10%
■ Multichannel sales projects—40%
■ New products and product improvement projects—50%

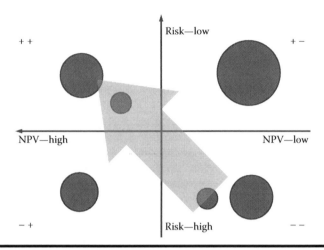

Figure 5.7 Company G: Clothing manufacturer portfolio balance—NPV vs. risk.

Summary

At the beginning of this chapter, we mentioned several challenges faced by the new product development companies around the world. They included

- Globalized competition
- Emergence of new markets, especially in India, China, Russia, and Brazil
- Need for innovation
- Ability to protect IP

A quick review of the companies' scoring models reveals that six out of seven of them included the strategic fit as one of the prioritization model variables, whereas in most of the cases, company strategies included the preparation for global competition, a focus on specific geographic markets, and a need for innovation.

Furthermore, five out of seven organizations decided to include the "synergies with the existing products" variable into their models. This could be interpreted as the preparation for global competition in an attempt to cut marketing costs and improve overall efficiency.

Finally, "market attractiveness" and "financial factor" were used by six out of seven organizations. Again, this could be explained as a desire to improve financially responsible innovation to be prepared for even strong competition on the global scale.

Chapter 6

Project Portfolio Management in the Financial Industry

Financial Sector Overview

The financial services industry has experienced a crisis comparable to the one that happened about 80 years ago, the sad period of time we have been referring to as "The Great Depression." The financial crisis of 2008–2012 had a major impact on the banks, insurance, and investment companies alike. Many of them went into "preserve" mode, attempting to salvage their profits, market shares, and customers. However, with a slow recovery looming at the end of the proverbial tunnel, the financial sector had to develop a new game plan for the future under more favorable economic conditions.

Let us try to analyze the reality of the modern financial industry. First, as was mentioned earlier, the financial crisis slowed the development of new products and services at the banks while they were trying to hold on to the "status quo."

Second, both international and national financial governing bodies started to introduce tougher regulatory legislation, which managed, according to some financial practitioners, to harness the creativity of the financial sector.

Because these two factors coupled with the decrease in the individual wealth of people, the competition for the share of wallet increased. Furthermore, certain markets have reached a near-saturation point where the entire customer base has already been divided among several key financial players.

What are the possible solutions for such an environment? If one uses a purely logical or even a mathematical model, several potential conclusions can be made.

The companies in the financial industry can expand their businesses and increase their profit margins by

1. Introducing existing products to the existing markets
2. Introducing new products to the existing markets
3. Introducing existing products to new markets
4. Introducing new products to the new markets

Obviously, the first proposition is the weakest one on the list, especially if we consider increased local competition, decreasing prices, and market saturation. On the other hand, the second, third, and fourth propositions offer, at least in theory, a way out of the existing situation (see Figure 6.1).

Any of the three strategies requires these companies to increase the number of products and services and/or penetrate new markets. This is easier said than done because of several challenges financial institutions face.

The ability to offer new products and services is highly correlated with the technological preparedness of the organizations; this notion includes aspects such as e-banking, mobile banking, and online trading platforms. Interestingly enough, 28% of the Deloitte survey participants indicated that their core banking IT systems need upgrades or complete replacement. In layman's terms, this implies about

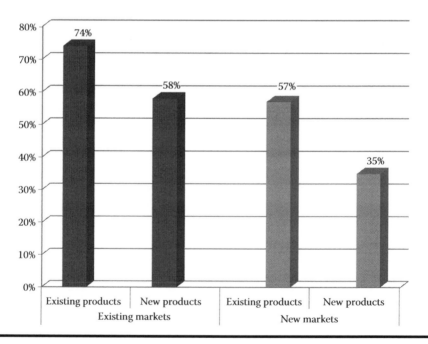

Figure 6.1   What growth strategies is your company pursuing?

one-third of the financial institutions. Rather than focusing on new modern applications that would run on top of the existing core systems, these companies will have to dedicate their time and resources to fixing the "IT heart" of their businesses.

Another issue is the much needed improvements to the customer service. This problem is threefold. On the one hand, customer service is typically the first area together with IT to face cuts during difficult economic times such as those we recently experienced. On the other hand, the introduction of new services and products requires the companies to train their customer-facing employees and provide them with additional expertise about the new offerings. Finally, if the organization decides to move into new markets, it will also need to upgrade its customer service with respect to language, legal, and sometimes even cultural trainings.

When entering the new markets, organizations frequently discover that the local players are not particularly happy about their intrusion and hence initiate the price wars, which also lead to stiffer competition and lower prices. This fact, frequently coupled with increased expenses on marketing, also applies downward pressure on the profit margins.

It is not surprising then that when asked "how has the competition in your industry changed in the last 12–18 months?" 72% of the financial sector executives in the Deloitte survey answered that the competition has increased (see Figure 6.2).

What conclusions can be made about the future of the financial sector and, more importantly, about the factors that will differentiate the true winners of this race? It looks like that the companies will need to

1. Introduce new products and services to attract new customers
2. Enter new markets
3. Upgrade their core and peripheral IT platforms
4. Somehow balance growth with profitability

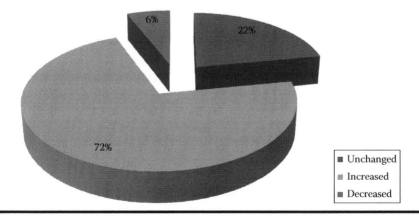

Figure 6.2 Competition growth in the financial industry.

5. Minimize the risks as much as possible, something that can present a challenge, especially considering points (1), (2), and (3)
6. Preserve their reputation
7. Comply with the stiffening and ever-changing regulatory rules

Interestingly enough when asked to name the areas that were deemed important for strategy, the executives in the financial sector named the following topics:

- Reputation—90%
- Return on investment (ROI)—87%
- Risk—79%
- Core competency—75%
- Regulatory compliance—62%

Financial Sector Case Studies

Introduction

The following section of the chapter will be dedicated to the presentation and the detailed analysis of three portfolio models developed by three different financial institutions. The first one is an eastern European bank that managed to weather the financial storm that ravaged the financial industry in the last several years relatively unscathed. The second organization is a western European banking institution that has experienced relative stagnation and even a slight dip in its revenues and profits. And finally, the third case is a North American brokerage firm that experienced serious regulatory challenges and was close to bankruptcy.

Let us examine and analyze their project portfolio models and assess how each one of the companies dealt with its unique situation.

Eastern European Bank A

The first case study examines a European bank providing an array of financial services in a medium-sized country. It provides both retail and corporate services including savings and current accounts.

The bank provides retail, corporate, and investment banking services in the Czech Republic. The company offers a range of banking and other financial services, such as savings and current accounts, asset management, consumer credit and mortgage lending, investment banking, securities and derivatives trading, portfolio management, project finance, foreign trade financing, corporate finance, capital and money market, and foreign exchange trading.

The bank is a fairly old institution founded in the nineteenth century and currently employs between 7,000 and 13,000 employees. At the time of the portfolio management exercise, the institution was the second-largest bank in its country (based on total assets) but was experiencing stiff competition from approximately ten other institutions operating in the country.

The organization was able to survive the financial crisis relatively unscathed; while it did not incur any significant losses, its net income remained stagnant over the course of the five years prior to 2014, while its total assets grew by only 1.6% per annum.

Strategy

The company managers wanted to focus on the following strategic goals:

- To become a number one bank, at least in certain segments including money markets, mergers and acquisitions, corporate banking, medium and small enterprises
- To improve customer satisfaction and gain a larger market share
- To increase its profit by 20% per year even though it had zero-rate growth in the previous five years
- To keep its cost-to-income ratio at less than 40%

Scoring Model

The scoring model creation exercise conducted with the company's executive team led to identify the following scoring criteria (see also Table 6.1):

- Strategic fit
- Net present value (NPV)
- Payback
- Execution risk

The senior management committee also decided to award the following points for the "high," "medium," and "low" categories:

- High—10 points
- Medium—5 points
- Low—1 point

The managers decided to award 1 point for a candidate project that fits only one of the strategic fit criteria, 5 points for the proposals that satisfy two criteria, and 10 points for the candidates satisfying three or four (i.e., all) of the strategic fit criteria. Furthermore, if the project proposal does not fit any of the strategic targets, it would automatically be removed from the pipeline without considering its other attributes.

Next, for the NPV category, it was decided that any project proposal that has an NPV of less than €X would get a rating of 1 point. Proposals with NPVs that fall

Table 6.1   Eastern European Bank A Portfolio Scoring Matrix

Selection Criteria	Points Awarded (Maximum Possible 40)			
	41 Points			
Joker	1 Point	5 Points	10 Points	Kill?
Strategic fit	Low Fits one of the criteria	Medium Fits two of the criteria	High Fits three or four of the criteria	Yes (if fits none of the strategies)
NPV	Low NPV < €X	Medium €X < NPV < €Y	High NPV > €Y	No
Payback	Long P > 3 years	Medium 1 < P < 3 years	Short P < 1 year	No
Execution risk	High Unknown technology New type of project	Medium Somewhat known technology Project scope somewhat familiar	Low Known technology Done similar projects in the past	No

anywhere between €X and €Y would get a rating of 5 points, whereas NPVs larger than €Y will get a score of 10 points.

This scoring criterion required a long discussion regarding what score would fall into the "kill" category. The initial inclination of the executives was to kill all the project proposals with an NPV less than zero (i.e., the projects that would lose rather than generate additional funds for the company). However, once the management team considered a large number of IT initiatives that required implementation, they changed their minds about this category. The problem with this organization, as well as with many others even outside the financial sector, is that IT plays a major role in its development. However, it is difficult to find an IT project that has a positive NPV. For example, if we replace the outdated core banking system, it would cost us millions of euros, and yet it will not generate any additional cash inflows. In other words, the bank would make exactly the same amount of money with either the old or the new system. But if we do not implement this project, there is a considerable risk that one day the entire company operations can come to a standstill.

The next category selected was the "payback"—the length of time required to recover the cost of an investment. The senior managers decided to award 1 point for projects providing a full payback in more than three years, 5 points to projects providing a payback in between one and three years, and 10 points to project proposals that promised to fully pay for their investments in one year or less.

Finally, the company leadership wanted to consider the execution risk of the upcoming projects. In other words, they wanted to assess the complexity of the proposed project. If the project was fairly complex, involving unknown technologies and those never done in the past, it would get a score of one. If the proposed venture was of medium complexity, it would receive 5 points, and if the proposed project involved known technologies and was similar to the projects done in the past, it would get 10 points.

Portfolio Balance

The executive team has decided to monitor the balance of the portfolio via two bubble charts (see Figures 6.3 and 6.4):

■ Payback vs. execution risk
■ NPV vs. execution risk

Strategic Alignment

Company management decided to proceed with a "top-down, bottom-up" model with strategic buckets. The executives designated the following buckets:

■ Maintenance or stay-in-business projects (30% of the resources measured in person-months)

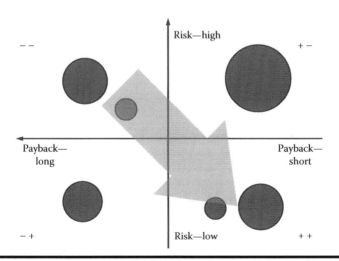

Figure 6.3   Eastern European bank A portfolio balance—payback vs. execution risk.

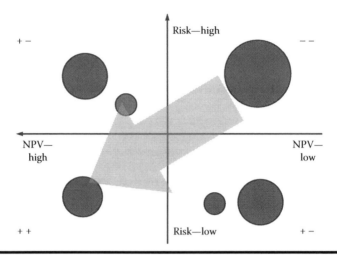

Figure 6.4 Eastern European bank A portfolio balance—NPV vs. execution risk.

- Business improvement projects (50% of the resources measured in person-months)
- Breakthrough projects (20% of the resources measured in person-months)

This project breakdown demonstrates what one may call a "moderately aggressive" approach to the project portfolio selection, which reflects the good standing of the bank and its desire to expand its business.

Western European Bank B

The second case study focuses on a western European subsidiary of a large multinational banking and financial services corporation. While the subsidiary operates only in a medium-sized European country, the parent organization is present in dozens of countries and serves millions of customers.

The subsidiary in question has managed to survive the financial crisis relatively unscathed, but it still had certain performance issues including stagnant income numbers and even a slight dip in 2012.

As a result of these issues, the senior management decided to analyze and prioritize their projects as well as to better align them with the company strategy for the next three to five years.

Strategy

Considering its previous challenges, the executive management team developed the following strategy:

- Offer online 50% of the future "simple" sales to reduce operating costs.
- Offer online 100% of future simple services, again, to reduce the operating costs.
- Improve transparency and understanding by offering all of the products and services using "easy to understand language."
- Ensure all products and services introduced by the global headquarters undergo product nationalization to conform to local laws and standards.
- Become a major employer in the country.

Scoring Model

The senior management team agreed on the following scoring model for company project proposals (see also Table 6.2):

- NPV
- Payback
- Strategic fit
- Technical project risk
- Customer impact (importance for customer)
- Employee impact (potential decrease in the headcount)

The first category in the model was the NPV of the proposed projects. Projects with an NPV less than €1 million would get a score of 1 point, the proposals with an NPV between €1 and €5 million would get a score of 5 points, and finally, ventures whose NPV promises to exceed €5 million would receive 10 points.

An interesting discussion happened regarding the "kill" designation for this category. Initially, the managers designated a proposed project with a negative NPV as a "kill" category that automatically would be removed from the list of projects without any further consideration. However, once the facilitator mentioned that this decision effectively removed all future IT upgrade projects, the decision was reconsidered. It was decided to designate this category as "kill" for all new product and service projects but keep it as a "no-kill" for IT maintenance ventures.

Payback was the second variable added to the model to promote projects that would fully recover their costs sooner rather than later.

Strategic fit was an obvious candidate for addition to the model. To impose some measurability on this category, the managers decided to award 1 point to proposals that fit at least one of the strategy criteria, 5 points to the ventures including between two and three of the strategic criteria, and 10 points to the projects including between four and five strategic priorities. This particular category has been designated as a "kill" for the projects that aligned with none of the strategies.

Technical project risk also was added to the variable mix to promote projects involving familiar technologies and to "penalize" ventures including unknown platforms and "know-hows."

Table 6.2   Western European Bank B Portfolio Scoring Matrix

Selection Criteria	Points Awarded (Maximum Possible 60)			
	61 Points			
Joker	1 Point	5 Points	10 Points	Kill?
NPV	NPV < €1 million	€1 million < NPV < €5 million	NPV > €5 million	Yes, if less than zero for new products and services No, for "stay-in-business" projects
Payback	P > 3 years	2 < P < 3 years	P < 2 years	No
Strategic fit	Fits one of the criteria	Fits two or three of the criteria	Fits four or five of the criteria	Yes, if does not fit any of the criteria
Technical risk	Known technology	Somewhat unknown technology	Completely unknown technology	No
Customer impact (market attractiveness)	Low Few customers require such product or service	Medium Some customers require such product or service	High Many customers require such product or service	No
Employee impact	Many employees may be laid off	Some employees may be laid off	No or few employees may be laid off	No

The fifth category, "customer impact," would measure the attractiveness of the new product or service to the customer base. After having a long debate regarding the measurability of this category, the decision was to use the following scheme:

- Low: Few customers require such product or service—1 point
- Medium: Some customers require such product or service—5 points
- High: Many customers require such product or service—10 points

Finally, the executives insisted on adding a fifth category they called "employee impact" since they were seriously concerned about any possible negative impact on the reputation of the bank. Therefore, project proposals that could potentially lead to significant layoffs would get a rating of 1 point, while projects leading to small and no layoffs would get 5 and 10 points, respectively.

All of the aforementioned implies that a project candidate in this model could get a maximum score of 60 points and a minimum of 6 points. The executives also discussed at length the "joker project" concept, that is, the project proposals that score low on a proposed model but may have a potential breakthrough impact on the company business. The executive team has decided to award a default 61 points to such proposals, thus taking them to the very top of the rank-ordered proposal list. In order for a project candidate to receive the "joker" rating, it had to be approved by the company's CEO.

Portfolio Balance

The company management team decided to monitor the portfolio balance using the technical risk vs. NPV bubble chart (see Figure 6.5).

Strategic Alignment

The management team decided to proceed with a "top-down, bottom-up" model with strategic buckets and designated the following buckets:

■ Maintenance or stay-in-business projects (20% of the resources measured in person-months)

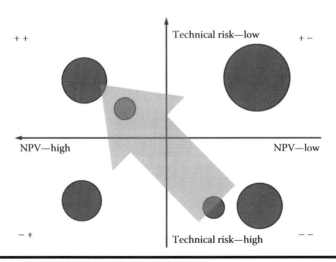

Figure 6.5   Western European bank B portfolio balance—technical risk vs. NPV.

■ Business improvement projects (50% of the resources measured in person-months)
■ Breakthrough projects (30% of the resources measured in person-months)

When compared with the eastern European bank A example, one can see that this organization has employed a more aggressive strategic bucket approach with 30% of the total project resources dedicated to potential breakthrough projects.

North American Brokerage Company C

The third company in this chapter is a North American financial brokerage company with a global presence in the United States, Canada, the United Kingdom, and several European countries. After several years of successful operations worldwide, the company received stiff fines and penalties by the Securities and Exchange Commission for what was referred to as "trading irregularities."

As a result of these events, the financial situation of the company worsened significantly with its stock plummeting to historical lows.

Consequently, the senior management of the firm decided to use project portfolio management to select the best projects that would potentially lead the company out of the financial troubles it faced and to prioritize them properly so that the projects selected received the resources required for their successful completion.

Strategy

The executives of the firm decided to concentrate on the following four pillars to guide the company out of the crisis:

1. Expand its fixed-income business to other regions outside the United States to generate additional revenues.
2. Consolidate equity franchises to one business to decrease operating costs.
3. Upgrade online retail offering to new platform, since the old one was outdated and had a multitude of performance issues.
4. Consolidate regional general ledgers into one global system to reduce operating costs.

Scoring Model

During a facilitated session, the executives identified the following scoring variables (see also Table 6.3):

■ Strategic fit
■ Revenue generation/cost avoidance

Table 6.3 North American Brokerage Company C Portfolio Scoring Matrix

Selection Criteria	Points Awarded (Maximum Possible 50)			
	51 Points			
Joker	1 Point	5 Points	10 Points	Kill?
Strategic fit	Fits one of the criteria	Fits two of the criteria	Fits three or four of the criteria	Yes (if does not fit any of the criteria)
Revenue	R < $5 million	$5 < R < $10 million	R > $10 million	No
Time to market	T > 16 weeks	7 < T < 15 weeks	T < 6 weeks	No
Project size and cost (resources)	R > 180 man-months	30 < R < 180 man-months	R < 30 man-months	No
Existing expertise (leverage of core competencies)	No existing expertise	Some existing expertise	Lot of existing expertise	No

■ Time to market
■ Project size and cost
■ Existing expertise (leverage core competencies)

For the strategic fit category, the managers decided to award 1 point for the proposals that fit only one of the strategic initiatives, 5 points for the ventures with two strategies, and 10 points to projects that included three to four of the strategies. Furthermore, the projects that did not fulfill any of the strategic initiatives were automatically "killed" without any further consideration.

Concerning the revenue aspect, they decided that projects that "promise" to generate less than $5 million would get 1 point, while projects generating between $5 and $10 million and more than 10 million would get 5 and 10 points, respectively. Despite the financial difficulties, the managers decided not to designate this category as "kill" since several of the company's projects mentioned in the strategy IT upgrade ventures with no positive revenues associated with them.

Considering the difficult financial situation the company found itself in, time to market was considered a very important criterion. As a result, the managers designated aggressive timelines for their scoring model:

- T > 16 weeks—1 point
- 7 < T < 15 weeks—5 points
- T < 6 weeks—10 points

An interesting discussion took place during the analysis of this variable. The facilitator noted that considering the grandiose initiatives mentioned in the company's strategy, it was unlikely that any of the major project proposals would score very high on the scale mentioned. However, the managers still insisted on imposing such tight timing requirements on all future projects. Can you add a why?

The managers also decided that project size would be measured in person-months required. Consequently, project proposals requiring more than 180 person-months of effort received 1 point; those needing between 30 and 180 person-months, 5 points; and the ones where effort would be less than 30 person-months, 10 points.

Finally, the category called "existing expertise" was designed to measure the company's internal ability to handle the proposed projects. The managers decided the easy projects were ones in which the company did not require external resources and would get 10 points. The projects where some external expertise would be required would get 5 points, and finally, the projects requiring extensive external involvement would get a rating of 1 point.

To sum up all the analysis, the maximum points a project proposal can get in this scoring model is 50 points and the minimum is 5 points. After a lot of discussion, the mangers agreed that they also wanted to use the "joker project concept" where a proposal that scores low in the model but has a chance to become a proverbial "breakthrough project" can receive a go if approved by the C-level committee.

Portfolio Balance

The bank managers decided to assess the balance of their project portfolio using the following bubble charts (see also Figures 6.6 and 6.7):

- Revenue vs. time to market
- Revenue vs. resources

Strategic Alignment

Considering the difficult situation the company found itself in, the executives, after a long discussion, decided to use a blend of the popular "top-down, bottom-up"

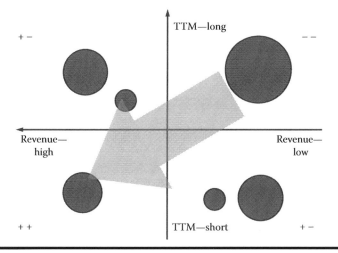

Figure 6.6   North American brokerage company C portfolio balance—revenue vs. time to market.

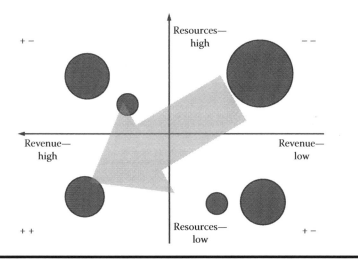

Figure 6.7   North American brokerage company C portfolio balance—revenue vs. resources.

approach with the more conservative "top-down" model. They recognized that the company had already listed most of its key projects in its strategy:

■ Expand the fixed-income business outside the United States.
■ Consolidate equity franchises.
■ Upgrade the online retail platform.
■ Consolidate the general ledgers.

The managers also decided that there would be some flexibility with respect to the "expansion of the fixed-income securities" project as they would be open to entertaining employee ideas as to where exactly this business should expand.

Eastern European Bank D

The financial organization in this case study is a fairly small private bank that also offers corporate and investment banking services to large- and medium-sized enterprises. At the time of the portfolio management workshop, the organization has been in existence for approximately eight years.

The shareholders of the bank expected the organization to achieve the leading role in the country's financial market and imposed fairly aggressive growth targets on the bank's management team. The situation was further complicated by the fact that the local market was heavily saturated, with approximately 20 competing banks vying for the same customers.

In addition, company management had serious concerns about the quality of their project portfolio and wanted to prioritize and—whenever possible—eliminate unnecessary projects.

Strategy

Based on this situation and the goals of the shareholders, the bank's strategy included the following components:

■ Increase the number of both small and medium enterprises and large business clients.
■ Increase the private customer base.
■ Increase portfolio diversification via new products.
■ Improve bad debt management.
■ Improve credit risk management.

Scoring Model

The scoring model developed as a result of the project portfolio management workshop consisted of the following six variables (see Table 6.4):

1. Strategic fit
2. ROI
3. Market attractiveness/competitive advantage
4. In-house expertise
5. Risk and complexity
6. Improve operational efficiency

Table 6.4 Eastern European Bank D Portfolio Scoring Matrix

Selection Criteria	Points Awarded (Maximum Possible 40)			
	41 Points			
Joker	1 Point	5 Points	10 Points	Kill?
Strategic fit How many of Bank's strategies do this project support?	Low one strategy	Medium two or three strategies	High four or more strategies	Yes If supports zero strategies
ROI What is the financial outlook for this project?	Low <10%	Medium 10%–12%	High 12+%	No
Market attractiveness/ competitive advantage How high is the market demand for this project? Are there many competitors offering same product or service?	Low Market demand is low. Many competitors offering similar products or services	Medium Market demand is medium. Some competitors offering similar products or services	High Market demand is high. Few or none of the competitors offering similar products or services	No
In-house expertise Does the bank have internal expertise to deliver this project? Will it require a lot of external resources?	Low The bank does not have internal experts. The project will require a lot of external expertise	Medium The bank has some internal experts. The project will require some external expertise	High The bank has a lot of internal experts. The project will require little or no external expertise	No

(*Continued*)

Table 6.4 (*Continued*)   Eastern European Bank D Portfolio Scoring Matrix

Selection Criteria	Points Awarded (Maximum Possible 40)			
	41 Points			
Joker	1 Point	5 Points	10 Points	Kill?
Risk and complexity Will this project encounter a lot of risks (e.g., technical, organizational, operational, legal, compliance, etc.) during its delivery? How complex is this project from the technical point of view?	High A lot of risks are expected to surface during the project. Very complex project	Medium Some risks are expected to surface during the project. Somewhat complex project	Low Few or no risks are expected to surface during the project. Simple project	Yes In case of legal or compliance risks
Improves operational efficiency? Will this project improve the operational efficiency at the bank? How significant is the expected impact?	Low Little or no expected positive impact on the operational efficiency	Medium Some expected positive impact on the operational efficiency	High Significant expected positive impact on the operational efficiency	No

The first variable added to the model was strategic fit. The points in this category were distributed as follows:

- 1 point—The project is aligned with only one strategy.
- 5 points—The project is aligned with two or three strategies.
- 10 points—The project is aligned with four or more strategies.

The second variable included was ROL. If the proposed project was expected to generate less than 10% in return, it would be given 1 point. If the venture was expected to generate between 10% and 12%, it would be awarded 5 points, and finally the projects expected to generate a return of more than 12% would get 10 points.

The third category was the combined market attractiveness/competitive advantage variable. The points were awarded in the following fashion:

- 1 point—Market demand is low. Many competitors offer similar products or services.
- 5 points—Market demand is medium. Some competitors offer similar products or services.
- 10 points—Market demand is high. Few or none of the competitors offer similar products or services.

The next variable added to the scoring matrix was the in-house expertise required to accomplish the project. The points were distributed in the following way:

- 1 point—The bank does not have internal experts. The project will require extensive external expertise.
- 5 points—The bank has some internal experts. The project will require some external expertise.
- 10 points—The bank has numerous internal experts. The project will require little or no external expertise.

The fifth category considered was the risk of the proposed project. If the venture carried a lot of associated delivery risks, it would get 1 point; if the project was deemed to carry a reasonable [such as] amount of risk, it would get 5 points; and finally, if the project was expected to carry only a few risks, it would be awarded full 10 points.

Considering the pressure from the shareholders to reduce costs, the management team decided to add the "improves operational efficiency" variable to the mix to promote the cost-saving projects. The points in this category were awarded in the following manner:

- 1 point—Little or no expected positive impact on the operational efficiency
- 5 points—Some expected positive impact on the operational efficiency
- 10 points—Significant expected positive impact on the operational efficiency

In addition, the managers decided to accept the "joker" project concept by awarding it to either regulatory projects or to ventures that scored low in the scoring matrix but were deemed important by the C-level executives. These endeavors would automatically receive a score of 61 points, thus taking them to the top of the rank-ordered project list.

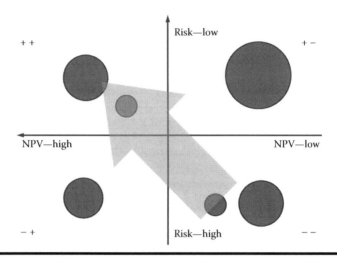

Figure 6.8   Eastern European bank D portfolio balance—NPV vs. risk.

Portfolio Balance

The company management team decided to monitor the portfolio balance using the risk vs. reward bubble chart (see Figure 6.8).

Strategic Alignment

The company management decided to adopt the "top-down, bottom-up" methodology with buckets as follows for the resource allocation:

- Maintenance projects—30% of financial resources
    - Cost of staying in business ventures
    - Risk management, compliance, other regulatory projects, projects endorsed by the government and shareholders
- Enhancement projects—40% of financial resources
    - Typically medium ROI/medium-risk venture
    - Improvements of existing strategic capabilities/product platforms
- Innovation projects—30% of financial resources
    - High ROI/high-risk ventures. Could be a source of outdistancing the competition

Summary

How can one summarize the four examples shown especially in the light of the outlook for the financial industry?

First, financial institutions must simply employ a combination of expansion into new territories and introduce new products to successfully compete and grow their businesses.

Also, upgrades to the existing IT core platforms as well as introduction of the new systems will play a major role in the future.

Furthermore, the banks must accomplish the aforementioned tasks while actively managing risks, profitability, reputation, and regulatory relationships.

All of the four companies analyzed used "strategic fit" as one of their variables in their scoring models, which by itself is not overly surprising. Having said that, the strategies "embedded" in these criteria were quite different depending on the situation at the organizations.

Again, it is not surprising that all four of the companies used some financial measure to assess their projects, with the first two institutions using NPV and payback, the third one using revenue (or costs avoided), and the fourth one using ROI.

Risk also was an important factor in all four models in various forms: the first company called it "executions risk"; the second, "technical project risk"; the third one, "existing expertise"; and the fourth one, "risk and complexity."

The more significant differences appeared later. For example, in the first case study, the bank was in good financial standing, hence deciding to concentrate more on strategic, financial, and risk aspects, while the western European bank, which had experienced certain difficulties, albeit minor, added customer impact (i.e., attractiveness) and employee impact to the mix.

The North American financial institution discussed in the third example focused on time to market and project size to address its difficult financial situation. And finally, the fourth organization decided to concentrate on smaller and simpler projects while improving their operational efficiency.

Chapter 7

Project Portfolio Management in the Energy and Logistics Industries

Overview of the Energy and Logistics Sectors

The energy industry has undergone profound changes in the last several years: historical highs for oil prices have recently been replaced with a significant drop along with the shale gas revolution coupled with the growth of alternative and renewable energy. According to noted economist and author Jeremy Rifkin (2011), we are entering the third industrial revolution, characterized by the effective convergence of the breakthrough energy and modern communications technologies.

As of 2014, renewable energy (hydro, biofuels, waste, geothermal, solar, wind, etc.) accounted for less than 20% of the global electricity generation (Deloitte, 2012a). The rest were as follows:

- Coal (dirtiest)—over 40%
- Natural gas—22%
- Nuclear—13%
- Crude oil—5%

To add to the chaos, oil prices have lost almost 50% at the end of 2014, thus affecting the revenues of oil-producing organizations. How should the energy companies respond to such volatile environment? According to leading energy industry experts (Deloitte, 2012b), there are several factors that the companies must consider.

One is the lack of financing and capital as a result of a significant drop in the price of oil. Many companies are experiencing a lack of financing and capital when initiating their R&D and exploration projects. In addition, the energy organizations focusing on the renewable sector must deal with the subsidy environment, as subsidies to fossil fuels in 2011 totaled $523 billion, 30% more than the year before and six times more than the renewable energy.

This situation will have at least two implications for the companies in this domain. First, energy companies must be frugal with their money. Second, they will be forced to be careful with their investments just to make sure they select the highest-value projects for their organizations.

Another concern is the intense competition. As the energy prices decrease, the competition is intensifying as the companies try to increase their sales by offering new products and services and by moving into new markets.

Further, these companies are faced with deregulation as another concern for monopolistic players in the field as well as for companies whose markets were protected by the local governments. Energy deregulation is gaining its momentum in Europe, North America, and some other regions. The companies that did not have to worry about the potential competition or only had several other small players in the market now are faced with aggressive large companies moving across the borders and offering their energy products and services at much lower prices.

Risks further increase as the energy prices drop, revenues decrease, and the competition intensifies, and energy companies must be careful in managing their project risks, both at the project inception (the business case stage) and during the actual project planning/execution phase.

Finally, because of several recent accidents that resulted in serious environment damage and led to major fines and penalties, energy companies must continue to invest in safe, reliable, and compliant operations.

Energy and Logistics Sector Case Studies

Introduction

This particular chapter provides examples of different portfolio models developed in my facilitation sessions with several energy companies around the world. The five companies to be discussed include a power trading agency, western European electric utility, regional IT department of a global oil and gas producer, eastern European electricity company, and an IT team of an energy operator also involved in global logistics.

Energy Company A: Power Trader

The first company discussed in this chapter is a large electricity trader with operations in several countries.

This organization was only several years old at the time of portfolio model creation and wanted to expand into several other countries while battling fierce competition in a deregulated market.

In addition, the executive management felt that the only way to reach its goals would be a disciplined and measurable approach to the project portfolio selection and prioritization to deliver the best products and services to the market.

Strategy

The company's strategy developed to address the issues and challenges it was currently facing consisted of the following five points:

1. Extend market to other neighboring countries.
2. Add more customers.
3. Actively participate in the continental energy market integration.
4. Actively develop new products.
5. Improve business processes aimed at decreasing risks and cutting costs.

Scoring Model

The scoring model developed by the company executives during the portfolio management workshop consisted of the following five variables (see Table 7.1):

1. Strategic fit
2. Competitive advantage
3. Market attractiveness
4. Technical feasibility
5. Financial (payback)

In the strategic fit category, the project received 1 point for addressing one of the company's strategies, 5 points for covering two to three strategies, and 10 points for covering all four of the strategic initiatives. If a project did not support any of the strategic initiatives at all, unless they were deemed to be government-mandated initiatives or "joker" projects, they were removed from the portfolio.

The second variable was the competitive advantage. The points' breakdown was distributed in the following manner:

- More than two competitors offering similar products—1 point
- Either one or two competitors offering similar products—5 points
- No competitors offering similar products—10 points

Table 7.1 Energy Company A: Power Trader Portfolio Scoring Matrix

Selection Criteria	Points Awarded			
	51 Points			
Joker	1 Point	5 Points	10 Points	Kill?
Strategic fit	Fits one of the strategies	Fits two or three of the strategies	Fits four or five of the strategies	Yes, if fits zero of the strategies and not a "joker" or regulatory project
Competitive advantage	Low More than two competitors offering similar products	Medium Between one and two competitors offering similar products	High No competitors offering similar products	No
Market attractiveness	Will decrease the market share	Will maintain the market share	Will grow the market share	No
Technical feasibility	Completely unknown domain	Somewhat familiar domain	Very familiar domain	Yes, if completely unknown and very risky and not a "joker" or regulatory project
Financial (payback)	P > 3 years	2 < P < 3 years	P < 2 years	No

Another category added to the prioritization model was the market attractiveness factor. The executives chose to correlate this variable with the degree of its influence on the market share. As a result, the project scoring model looked as follows:

- The proposed project will decrease the market share—1 point
- The proposed project will maintain the market share—5 points
- The proposed project will increase the market share—10 points

The technical feasibility of the proposed endeavor was also something the senior management wanted to consider when assessing the project value, as they wanted to move away from larger, more complex projects requiring a lot of external—and frequently expensive—expertise. As a result, the projects requiring extensive expertise would receive 1 point, the projects where the domain was somewhat familiar to the internal company employees would get 5 points, and the ones where the entire project domain was completely familiar would receive 10 points.

In addition, the executives designated projects as a "kill" variable for the exceptional cases where the project knowledge was completely unknown, and the project was deemed to be risky.

Finally, the financial factor in the form of a payback has been added to the portfolio scoring model. The breakdown of the points looked as follows:

- P > 3 years—1 point
- 2 < P < 3 years—5 points
- P < 2 years—10 points

Portfolio Balance

The senior managers decided to monitor the portfolio balance by periodic examination of the risk vs. reward bubble chart of their projects' portfolio (see Figure 7.1).

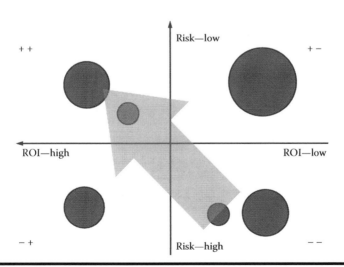

Figure 7.1   Energy company A: Power trader portfolio balance—ROI vs. risk.

Strategic Alignment

The executive committee decided to adopt the classical strategic alignment model and break down the portfolio into the following strategic buckets:

- Maintenance or utility projects: Support ongoing products and services—20%
- Growth or enhancement projects: Support strategic initiatives and increase value—60%
- Transformation projects: The new products or services that will hopefully dominate the marketplace—20%

Energy Company B: European Electric Utility

The second energy company discussed in this chapter is a large European utility organization that has been formed through a series of mergers in the past several years. The main area of operations of the company is production and sales of the electricity to several large European countries.

In the several years prior to the portfolio management initiative, the company started to experience certain challenges, mainly ones that were due to the changing environment in the market. They included mainly downward pressure on the revenues because of deregulation and increased competition.

Strategy

As a result of the problems and challenges the company faced, the executive management came up with the following strategy:

- Reduce costs and improve operational efficiency.
- Develop internationally (primarily outside of Europe).
- Keep its "A" credit rating.
- Develop synergies across activities (i.e., cross-sell).
- Create new services and find new clients.

Scoring Model

The senior management, after a long discussion, decided to go ahead with a seven-variable scoring model (see Table 7.2) as follows:

1. Strategic fit
2. Market attractiveness
3. Competitive advantage
4. Leverage core competencies
5. Net present value (NPV)
6. Payback
7. Commercial risk

Table 7.2 **Energy Company B: European Electric Utility Portfolio Scoring Matrix**

Selection Criteria	Points Awarded (Maximum Possible 70)			
	71 Points			
Joker	1 Point	5 Points	15 Points	Kill?
Strategic fit	Fits one of the criteria	Fits two or three of the criteria	Fits four or five of the criteria	Yes, if score is zero and not a regulatory or "joker" project
Market attractiveness	Low The new product or service will be used just by a few customers	Medium The new product or service will be used by some customers	High The new product or service will be used by many customers	No
Competitive advantage	Four or more of the competitors are offering a similar service	Two to three of the competitors are offering a similar service	One or less competitors are offering a similar service	No
Leverage of core competencies	Very little in-house expertise	Some in-house expertise	All expertise is in-house	Yes, if no in-house expertise at all and not a regulatory or "joker" project
NPV	NPV < €10 million	€10 million < NPV < €50 million	NPV > €50 million	No
Payback	P > 8 years	4 < P < 8 years	P < 4 years	Yes, if P > 15 and not a regulatory or "joker" project
Commercial risk	High	Medium	Low	No

The first category selected was the strategic fit of the proposed project. The executives decided to grant 1 point to the projects supporting one of their strategic initiatives, 5 points to the projects supporting two or three of the initiatives, and 10 points to the projects supporting more than four of the strategic initiatives. This variable was designated as a "kill" category unless the project was required because of the nature of regulations or was designated as a "joker."

The second variable added to the mix was market attractiveness. The points were distributed in the following manner:

- The new product or service will be used just by a few customers—1 point
- The new product or service will be used by some customers—5 points
- The new product or service will be used by many customers—10 points

The executives set up the next variable, competitive advantage, with the following measurable ranges:

- Four or more competitors are offering a similar service—1 point
- Two or three competitors are offering a similar service—5 points
- One or no competitors are offering a similar service—10 points

The fourth variable added to the model was leveraging core competencies. The managers decided to award 1 point to the project that required extensive external expertise, 5 points to projects where some external expertise was required, and 10 points to the projects that could be done in-house in their entirety. This variable had a "kill" category for any projects requiring more than 90% of external effort unless they were regulatory or "joker" initiatives.

NPV and payback were the next two variables included in the model with the points being distributed in the following fashion:

- NPV < €10 million—1 point
- €10 million < NPV < €50 million—5 points
- NPV > €50 million—10 points
- P > 8 years—1 point
- 4 < P < 8 years—5 points
- P < 4 years—10 points

The payback category also was designated as a "kill" variable for projects with the payback exceeding 15 years, again unless they were government-mandated or "joker" initiatives.

Finally, the final variable included was the commercial risk with the points distributed in the following manner:

- High commercial risk—1 point
- Medium commercial risk—5 points
- Low commercial risk—10 points

Portfolio Balance

The managers decided to monitor the company's portfolio status using the following bubble charts (see Figures 7.2 and 7.3):

- NPV vs. leverage of core competencies
- NPV vs. commercial risk

Strategic Alignment

The managers opted to proceed with the "top-down, bottom-up" approach to the portfolio alignment and chose to designate the following strategic buckets:

- "Stay-in-business" projects—10%–20%
- "Improve existing products and services" projects—50%–70%
- "The next breakthrough" projects—10%–40%

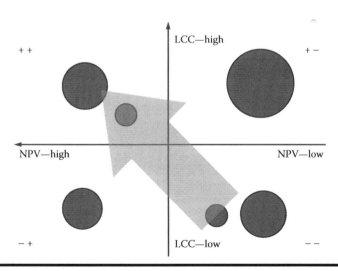

Figure 7.2   Energy company B: European electric utility portfolio balance—NPV vs. leverage of core competencies.

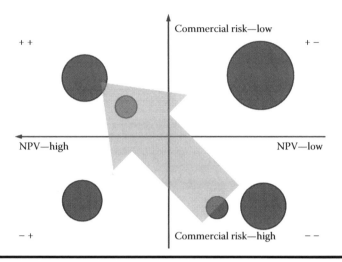

Figure 7.3 Energy company B: European electric utility portfolio balance—NPV vs. commercial risk.

Energy Company C: Regional IT Department of a Global Oil and Gas Producer

The company discussed in this section is one of the largest oil and gas producers in the world. In this particular case, we will examine the portfolio management system designed by one of its regional IT departments.

The situation at the company was such that all of the major IT projects were undertaken by the company headquarters, while the local IT departments were responsible mainly for servicing the needs of the offshore platforms. The executives of the regional department were under constant pressure as many of the projects they proposed were denied by the headquarters executives and yet they remained responsible for the safety, reliability, and security of all the offshore operations.

As a result, they felt that creating a portfolio scoring model would help them with (1) prioritization of their project proposals and (2) demonstration of the importance of their initiatives to the executive managers at the headquarters.

Strategy

The overall company strategy was developed at the organizational headquarters and consisted of approximately 10 strategic initiatives. However, the strategies directly related to the regional offices were

- Safety and reliability of all the operations
- Fiscal responsibility
- Simpler and more standardized procedures

Scoring Model

The scoring model created as a result of a one-day facilitated project portfolio management session is presented in Table 7.3.

As can been seen, it was an unusual model in comparison to the other scoring matrices described in the book. One may call it a purely risk-based approach to project prioritization.

The model included the following variables:

- Age of the technology platform or system—The project can be awarded between 1 and 5 points based on the age of the system.
- Business implications of the risk—The proposal received 1 point if the system failure would disrupt noncore (minor, internal) company operations, 3 points if it would disrupt external-facing company operations, and 5 points for the potential disruption of the site (platform) operations.
- Platform or system supportability—The initiative would receive between 1 and 5 points depending on the degree of vendor support.
- Platform or system intricacy—The project could be awarded 1, 3, or 5 points for a system serving a few, several, or many business units.
- Historical probability of failure—Again, the proposal could be awarded 1, 3, or 5 points for low, medium, or high (define these) historical probability of failure of such systems in the past.
- Risk register—Finally, the project received either 1 or 5 points depending on whether it was added to the company-wide risk register.

Therefore, a project proposal could receive between 6 and 30 points allowing a quick prioritization of the initiatives. In addition, the executives decided to designate the following ranges for the points awarded:

- High risk = 22–30 points—A "must do" project category; projects that must be initiated immediately
- Medium risk = 14–21 points—A "should do" project category; projects that should be approved or in special circumstances deferred for at most a year
- Low risk = 6–13 points—A "nice to do" project category; projects that can be postponed by two or three years and revisited at that time

Finally, the scoring model developed did not have a "joker" project category simply because the regional managers did not have the authority to unilaterally approve and initiate projects.

Portfolio Balance

No portfolio balance requirements were imposed on the model in question, mainly since all of the projects run by the local IT department would fall into the low-risk, low-reward category.

Table 7.3 Energy Company C: Global Oil and Gas Producer Portfolio Scoring Matrix

Selection Criteria	Scale	Description
Age of the technology platform or system How old is the current platform or system?	1, 2, 3, 4, 5	1 = 1 year 2 = 2 years 3 = 3 years 4 = 4 years 5 = 5 years
Business implications of the risk What is the impact of a system outage on business operations?	1, 3, 5	1 = Would disrupt noncore (minor, internal) company operations 3 = Would disrupt external-facing company operations 5 = Would disrupt site operations
Platform or system supportability Is vendor support available and to what degree?	1, 2, 3, 4, 5	1 = H/W is generally available and supported by the vendor 2 = H/W is no longer available but supported by the vendor 3 = Platform or system retirement is announced, but support is currently available 4 = No longer supported by the vendor (no bugs or patches are available) 5 = No support is available; spares are in short supply
Platform or system intricacy How many departments are impacted by the platform or system?	1, 3, 5	1 = Serves few business units 3 = Serves several business units 5 = Serves many business units
Historical probability of failure How likely is the failure of the platform or system?	1, 3, 5	1 = Low probability of failure 3 = Medium (normal) probability of failure 5 = High probability of failure
Risk register Has this proposal been added to the risk register?	1, 5	1 = No 5 = Yes
Points awarded High risk = 22–30 Medium risk = 14–21 Low risk = 6–13		

Strategic Alignment

Since this particular risk-based selection and prioritization model had direct linkage to the safety and fiscal responsibility initiatives outlined in the strategy section, and all of the projects would have fallen into the "maintenance" project category, the executives decided not to designate any special buckets.

It was mentioned at the end of the exercise that when several years of data are accumulated, it might be interesting to examine the project breakdown by the platforms.

Energy Company D: Eastern European Electricity Company

The energy company analyzed in this section is an eastern European electrical company that has until recently enjoyed a full monopoly, selling the electricity at one fixed rate regardless of whether it was dealing with private residences; small, medium, or large businesses; or government agencies. However, recent government legislation in the country led to the deregulation of the electricity market. This change implied that any energy company from three or four neighboring countries would be able to enter the market and compete with the former monopolist when it came to selling electricity to both private residences and businesses.

In addition, the company management felt that the value of the projects it had been delivering so far was too low. Also, the executives mentioned that they seemed to have too many initiatives under way while utterly lacking the resources (primarily human) to deliver all of them on time and on budget.

Strategy

The company's strategy was defined before the project portfolio workshop and, considering the recent deregulation, consisted of the following elements:

- Need to design attractive products. This implies
  - Various sizes of electricity packages
  - Fixed and variable rate packages to suit different customer needs
  - Extend loans to the customers needing them, especially the start-up businesses
  - Create different packages for households and businesses
- Increase revenues and profitability
- Improve public relations (PRs) damaged by the years of monopolistic presence in the market
- Social responsibility—initiate more green programs

Scoring Model

The scoring model developed during the facilitated portfolio management session contained the following variables (see Table 7.4):

- Strategic fit
- Competitive advantage
- Market share increase
- Time to break even
- Resources
- Technical complexity

The project proposal would receive 1 point if it fits only one of the four strategies, 5 points if it covered between two and three strategic initiatives, and 15 points if it absorbed all four strategies. This category has been designated as a "kill" for the projects that were not deemed regulatory or "joker" for those projects supporting zero strategies.

The second variable included in the model was competitive advantage. The managers decided to distribute the points in the following manner:

- More than three competitors in the area offering a similar product—1 point
- Two or three competitors in the area offering a similar product—5 points
- Zero or only one competitor in the area offering a similar product—15 points

The third variable added was the potential impact of the project on the market share increase (decrease), since the company was expecting a potential drop in the control of the market share because of the new deregulation laws. If the project was expected to increase the market share by between 0% and 1%, it would be awarded 1 point; if it was expected to increase the share by 2%–3%, 5 points; and finally, if it was expected to increase the market share by more than 3%, 15% points. Any projects expected to diminish the market share were as a "kill" variable and were automatically removed from further consideration unless they were regulatory or "joker" projects.

Time to break even was the fourth variable added to the model with points being distributed in the following manner:

- T > 3 years—1 point
- 1 year < T < 3 years—5 points
- T < 1 year—15 points

The fifth component was the resource requirements needed for the project. In the opinion of the executives, it served a dual role. First, it directly tied to the "grow revenues and profits" strategy, and second, it allowed the company to shift away from larger, more complex endeavors toward smaller "quick wins."

Table 7.4   Energy Company D: Eastern European Electricity Company Portfolio Scoring Matrix

Selection Criteria	Points Awarded (Maximum Possible 90)			
	91 Points			
Joker	1 Point	10 Points	15 Points	Kill?
Strategic fit	Low Fits one of the criteria	Medium Fits two or three of the criteria	High Fits four of the criteria	Yes, if fits zero of the strategic criteria unless a regulatory or "joker" project
Competitive advantage	Low More than three competitors in the area offering similar product	Medium Two to three competitors in the area offering similar product	High Zero to one competitors in the area offering similar product	No
Market share increase	Small 0%–1%	Medium 2%–3%	Large 3+%	Yes, if market size decreases, unless a regulatory or "joker" project
Time to B/E	Long T > 3 years	Medium 1 year < T < 3 years	Short T < 1 year	No
Resources	High Cost > $100,000	Medium $25,000 < cost < $100,000	Low Cost < $25,000	No
Technical complexity	Very difficult A significant external expertise will be required	Somewhat difficult Will need some external expertise	Easy Can be implemented by internal employees	No

As a result, projects costing more than $100,000 would receive 1 point; those with their budgets in between $25,000 and $100,000, 5 points; and the ones with the budget of less than $25,000, 15 points.

Finally, the executives of the company decided to add the "technical complexity" variable to the mix to penalize complex initiatives requiring an involvement of extensive external resources. The points were allocated according to the following scheme:

- A significant external expertise will be required—1 point
- Will need some external expertise—5 points
- Can be implemented by internal employees—15 points

Therefore, in the model created during the portfolio management session, the maximum number of points a project could get was 90, while the minimum was 6 points. If the project was of a regulatory nature or mandated by law, it would get an automatic score of 91 points, thus taking it to the very top of the prioritization list. The same procedure would apply to "joker" projects approved by the portfolio steering committee.

Portfolio Balance

The company's senior managers decided to monitor the portfolio balance using the following bubble chart models (see also Figures 7.4 through 7.6):

- Risk vs. time to break even
- PR risk vs. time to break even
- Technical complexity vs. market attractiveness

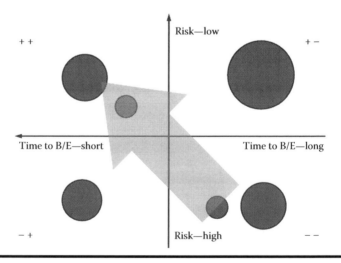

Figure 7.4 Energy company D: Eastern European electricity company portfolio balance—risk vs. time to break even.

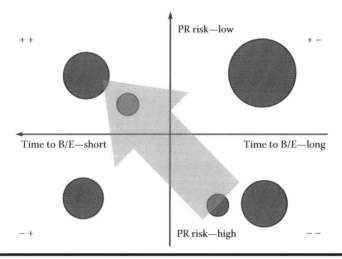

Figure 7.5 Energy company D: Eastern European electricity company portfolio balance—PR risk vs. time to break even.

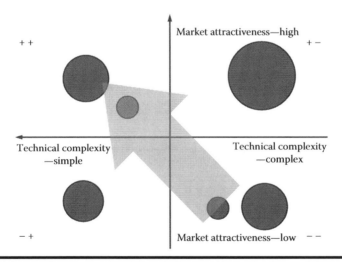

Figure 7.6 Energy company D: Eastern European electricity company portfolio balance—technical complexity vs. market attractiveness.

Strategic Alignment

The company decided to use an unusual two-pronged approach for the strategic alignment. The larger, strategic flagship projects—most likely the "jokers"—would be allocated via the top-down approach, while all the rest of the initiatives would undergo a top-down, bottom-up procedure.

The buckets designated for the top-down, bottom-up process were

- Maintenance
- Improvements to the existing products and services
- New products

Logistics and Energy Company A: The IT Team

The fifth company discussed in this chapter is the IT department of a global logistics company operating in North and South America, Europe, Asia, and the Middle East. During the conversation with the senior managers, the following problems were discussed:

- We have too many projects to implement.
- We do not have nearly enough resources.
- We constantly feel pressure to accept all of the project requests.

As a result, they felt that they needed to prioritize their projects in such a way so the value of the initiatives was the highest for the company and the size of the portfolio correlated with the throughput capacity of their department.

Strategy

The overall company strategy for the next five-year period consisted of the following six components:

1. Increase the return on investment (ROI) to 9%–11%.
2. Reach the earnings before interest and tax 6% mark.
3. Maintain or increase the market share.
4. Improve the operational efficiency (decrease costs).
5. Optimize the vessel network.
6. Improve customer care.

Scoring Model

The project portfolio scoring model created by the company's executives contained the following variables (see Table 7.5):

- Strategic fit
- Total cost of ownership (TCO) per year
- Size and complexity
- Risks

Table 7.5 Logistics and Energy Company A Portfolio Scoring Matrix

Selection Criteria	*Points Awarded*			
	61 Points			
Joker	*1 Point*	*4 Points*	*10 Points*	*Kill?*
Strategic fit	Fits one of the strategies	Fits two of the strategies	Fits three or more of the strategies	No
Total cost of ownership per year	TCO > €150,000	50,000 < TCO < €150,000	TCO < €50,000	No
Size and complexity	S > 50 man-months	15 < S < 50 man-months	S < 15 man-months	No
Risks	High Very risky project A lot of external expertise is required	Medium Somewhat risky project Significant external expertise is required	Low Low risks Almost no external expertise is required	Yes, if risks are very high, unless a regulatory or a "joker" project
Dependencies on other departments	High More than four other departments involved	Medium Three to four other departments involved	Low Zero to two other departments involved	No
Financial (NPV)	NPV < €20 million	€20 < NPV < €50 million	NPV > €50 million	No

- Dependencies on other departments
- Financial (NPV)

The first category added to the model was the strategic fit of the proposed project. The points were distributed in the following manner:

- The project fits one strategy—1 point
- The project fits two strategies—4 points
- The project fits three or more strategies—10 points

The next variable was the TCO per annum. The executives decided to add this variable to decrease the company's operational costs. They decided that projects delivering products or services with a TCO exceeding €150,000/year should be awarded 1 point; the ones with a TCO between €50,000 and €150,000, 4 points; and with a TCO less than €50,000, 10 points.

Project size and complexity was other factors added to the model, as the senior managers felt that they were not concentrating their resources on large, complex initiatives. The points were awarded in the following manner:

- Size > 50 person-months—1 point
- 15 < size < 50 person-months—4 points
- Size < 15 person-months—10 points

Project risks were the fourth variable added to the mix since the executives felt that they needed to improve portfolio risk management after several troubled projects they had.

If the project was deemed to be risky with a lot of external expertise required, it would get 1 point in the scoring model. If the project was one with average risk and needed some external expertise, it would be awarded 4 points. Finally, if the initiative carried little risks and required little external involvement, it would get a score of 10 points. Furthermore, the managers designated this variable as a "kill" category for the projects carrying extremely high risk.

Another interesting variable added to the model was the number of departments involved in the project. The executives felt that the project complexity tended to correlate with the number of business units engaged on the project and thus decided to include that variable to decrease the number of complex projects in the company's portfolio. The points were awarded in the following manner:

- More than four other departments involved—1 point
- Three or four other departments involved—4 points
- Zero to two other departments involved—10 points

Finally, the last variable added to the mix was the NPV of the proposed project. The points were distributed as follows:

- NPV < €20 million—1 point
- €20 < NPV < €50 million—4 points
- NPV > €50 million—10 points

As a result of this model, the project proposal could generate at least 1 point (unless it was "killed") and at most 60 points. If the project was deemed to be of regulatory nature or was awarded the "joker" status, it would get an automatic score of 61 points and move to the very top of the project prioritization list.

Project Analysis

We were able to run several past and current projects of the company through the newly created scoring model. The project list included

- Ship security system enhancement
- Company-wide enterprise resource planning (ERP) implementation
- Ship scheduling system implementation
- Cargo execution system implementation

The results of the scoring exercise are presented in Table 7.6.

The first project analyzed was the "ship security system enhancement" that was designed to upgrade all of the ship's security systems to provide defense against pirate attacks and possible hijackings. The project received 4 out of 10 possible points on the strategy fit since it covered the "improve the operational efficiency" and "improve customer care" initiatives.

In the "total cost of ownership," the project received all 10 points since the annual operating cost was less than €50,000 per annum. Furthermore, the project was granted 4 points in the "size and complexity" category since its size was expected to be somewhere between 30 and 40 person-months.

In the "risk" category, it was granted all 10 points because it was a low-risk project that could be accomplished by the internal resources.

There were at least three other departments to be involved in this endeavor, so the project was given 4 points. Finally, since the executives did not expect the project to generate any additional revenues, it was awarded 1 point in the NPV category.

Table 7.6   Logistics and Energy Company A Project Analysis

	Ship Security	ERP	Ship Scheduling	Cargo Execution
Strategic fit	4	1	4	1
Total cost of ownership per year	10	1	1	4
Size and complexity	4	1	1	4
Risks	10	1	1	4
Dependencies on other departments	4	1	1	4
Financial (NPV)	1	10	10	4
Total	33/60	15/60	18/60	21/60

Thus, the total number of points the project has generated was 33 out of a possible 60.

The second project discussed was the ERP system implementation. The project received the lowest possible scores in the first five categories because of low strategic fit, high TCO per year enormous size and complexity, high risks, and involvement of practically every department in the company. The executives insisted, however, that the project should generate extensive savings in excess of hundreds of millions of Euros, and as a result, they granted it 10 points in the NPV category. Regardless, the project received only 15 points in total, thus dropping it to the bottom of the prioritization list, it was considered as a "do, or go out of business" initiative, and the senior managers used their "joker" privilege and award this proposal 61 points.

The third project discussed was the "ship scheduling system." This initiative received 4 points in the strategy category since it was expected to address the "improve operational efficiency" and the "optimize vessel network" initiatives.

In the next four categories, it was awarded the minimum points available because of the high costs of ownership, large size, higher risks, and a number of dependencies on other departments. The executives felt, however, that the financial impact of this project would be significant, thus granting it 10 full points in the NPV category. As a result, the project received 18 out of 60 possible points.

Finally, the executive committee, with the help of the facilitator, assessed the "cargo execution" project that was supposed to enhance the loading and offloading of the freight. This particular project received only one point in the strategic fit category because it addressed only the "improve operational efficiency" initiative. In all the rest of the categories, it received 4 points each since it was perceived to be an endeavor of average cost of ownership, size and complexity, risks, dependencies on other departments, and financial benefits.

The total score of the project was calculated to be 21 out of 60 possible points.

Portfolio Balance

The company managers decided to monitor the portfolio balance using the following bubble chart (see Figure 7.7):

- NPV vs. strategic fit

Strategic Alignment

The portfolio committee opted for the popular top-down, bottom-down strategic alignment model with the following resource buckets:

- Regulatory projects—As required
- "Joker" projects—As required

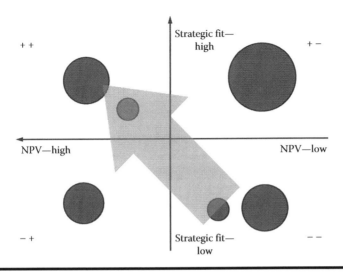

Figure 7.7 Logistics and energy company A portfolio balance—NPV vs. strategic fit.

- Maintenance ("stay-in-business") projects—20% of the resources remaining after regulatory and "joker" projects
- Efficiency improvement projects—50% of the resources remaining after regulatory and "joker" projects
- Breakthrough projects—30% of the resources remaining after regulatory and "joker" projects

Summary

At the beginning of this chapter, we stated that the energy market has been going through some profound changes requiring the companies to consider the following realities:

- They must be frugal with their money.
- They should be careful with their investments.
- They should be offering new products and services and moving into new markets.
- They have to be careful with managing their project risks, both at the project inception (the business case stage) and during the actual project planning/execution phase.
- Safety and compliance remain one of their top concerns.

Let us examine the strategic and portfolio models built by the five companies we examined in this chapter.

The strategy of the first organization, the power trading company, included elements such as extending its markets to other countries, adding more products and services, adding more customers, and improving the business processes. In addition, its portfolio scoring model had variables such as assessing the competitive advantage and market attractiveness to address the new products and service requirements, analysis of the technical feasibility to address the risk requirements, and, finally, the payback assessment to cover the fiscal responsibility prerequisite.

The European electric utility also had factors such as reducing costs and improving the operational efficiency, developing internationally, and creating new services in its strategy. Furthermore, its scoring model had variables such as competitive advantage and market attractiveness (new products and services), leverage of core competencies and commercial risk, and NPV along with payback to cover the financial responsibility demands.

The unusual portfolio model developed by the regional IT department of a global oil and gas producer was almost entirely risk and safety based, while its strategy included factors such as reliability of the operations and fiscal responsibility.

The eastern European energy company facing deregulation of the market added strategic initiatives to design attractive products and to increase revenues and profitability. Furthermore, the scoring model had variables such as competitive advantage (new products and services), market share increase (capture new markets and customers), time to break even and resources required (fiscal responsibility), and technical complexity (risk management).

Finally, the IT department of the logistics and energy company developed a strategy that was based on the financial performance and increasing the market share. In addition, its portfolio selection model was based on two major dimensions: risks (project size and complexity, technical risks, and dependencies on other departments) and financial performance (NPV).

Chapter 8

Project Portfolio Management in the Telecommunications Industry

Telecommunications Sector Overview

The worldwide telecommunications industry revenue is expected to reach $2.1 trillion in 2015 according to the market research firm Insight Research Corp. Despite the rocky global economy, industry revenue will grow further at an average annual rate of 5.3% to $2.7 trillion in 2017. The Asian region is seen as a key market and wireless revenue in this region is expected to grow to 64% (Insight Research Corporation, 2015).

What are the key factors that will influence the telecommunications sector in the next several years? According to various researchers, there will be several major changes affecting the worldwide telecom industry:

- Long-term evolution (LTE) or 4G adoption—An introduction of a new standard for wireless communication of high-speed data for mobile phones and data terminals.
- Improved and extended data service packages—Customers nowadays are increasing their data usage and demanding both higher-quality and cheap data usage rates.

- All you can app (AYCA) services—Many providers are offering the so-called all you can app services to their clients where, for a fixed fees, customers can enjoy unlimited usage of a package of apps.
- Major growth in developing markets—A significant growth in demand for wireless services (both voice and data) is expected in the developing countries.
- Mobile number portability (MNP) legislation—With many countries gradually adopting the MNP laws, allowing customers to switch freely between the mobile providers while preserving their phone numbers, the competition is expected to greatly increase.

The 4G adoption wave had a major impact on the telecom industry. It is expected that approximately 200 operators in 75 countries will switch to 4G by the end of 2013. The number of LTE subscribers has increased 17-fold between 2011 and 2013. Interestingly enough, prepaid markets (such as Russia, India, and especially China) have enjoyed faster LTE growth when compared to developed countries (Deloitte, 2013).

There are several important factors each company must assess before embarking on LTE implementation. First, these projects tend to be fairly costly, requiring major investments of both finances and human resources into technology upgrades.

It is also expected that wireless service providers would be willing to spend a lot of money on marketing of the new services as well as be prepared to offer their customers competitive pricing. Both of these factors may exert a downward pressure on the organizational bottom lines.

The battle for data seems to be yet another hurdle that many of the telecom companies have to overcome. In the United States, two of the carriers have started offering multidevice shared data plans for their customers. This implies customers purchasing a volume of data to be downloaded for a fixed price and then utilizing them according to their needs: for their smartphone, PC, laptop, or tablet. This trend is definitely a red flag for the companies offering broadband Internet connection services as their market share may start to shrink as customers would start to unsubscribe from their services. On the other hand, deployment of this service could also present problems to the telecom operators providing both mobile phone and broadband Internet services. What can take place in this situation is known as "cannibalization effect" in marketing, where customers would cancel their broadband services to switch to the universal data plans, thus having a zero net effect on the company as a whole.

A significant number of mobile operators around the world—between 50 and 100 according to estimates—are starting to offer an AYCA service where customers are provided with unlimited access to a certain package of applications for a fixed fee. Some experts think that this new service may act as a catalyst for an increase in a number of smartphones purchased especially in the developing world.

The number of countries who introduced the so-called mobile number portability laws is rapidly growing. This development presents the companies with a multitude of challenges, including keeping the existing clients and acquiring new customers by lowering their voice and data fees, offering new services and products, and improving the quality of their networks as well as customer service. For example, according to Delloitte, the top four reasons why customers decided to switch their mobile providers are as follows:

1. Price of voice and SMS tariffs
2. Quality of network coverage for voice calls
3. Customer support on the phone
4. Price of internet tariffs

So what are the strategic priorities of the telecommunications companies for the next several years? The first priority is investments into technology upgrades to move from 3G to 4G and to improve the quality of the Internet connection. Second, the companies will need to invest a lot of money into aggressive marketing of their new product and services, especially the transition to LTE. Third, the ability of the organizations to create new and innovative products and services will also play a major role in their survival. And finally, yet other important factors include competitive pricing and improved customer service.

Telecommunications Sector Case Studies

Introduction

In this chapter, we will examine four different telecommunications organizations. The first one is a dominant player in a local market that is facing new aggressive competition as well as continuing deregulation of the mobile phone marketplace. The second company is a proverbial underdog player with full of hopes and aspirations that could be marred by network quality issues.

The third example is an organization that is leading by innovation in a fiercely competitive market that is trying to keep its leadership position. And finally, the fourth case study describes a mobile provider that has experienced recent financial issues, operating in a fairly small market with very aggressive deregulation laws.

This chapter will describe the portfolio models these companies developed and attempt to link and explain their strategic choices with respect to both internal and external factors influencing the organizations.

Eastern European Mobile Provider A

This particular company has enjoyed a very good decade in an eastern European country. In the course of 10 years, it managed to get more than 50% of the market and posted great financial results year after year.

However in the past several years, the status of the organization has been seriously challenged by several other mobile services providers in the market: aggressive marketing, cheaper voice and data plans, and strong financial position of the competitors started to gradually chip away the company sales.

In addition, the government of the country has implemented an MNP legislation that allowed subscribers to freely change their mobile providers while keeping their original phone numbers.

These events, coupled with a major increase in the number and the complexity of their projects, forced the management of the company to initiate a serious discussion regarding the company strategy and project portfolio.

Strategy

The long-term strategy developed in the first facilitated session was aimed to define the top priorities of the management team for the next three- to five-year period. The major goals of the company turned out to be the following:

- Improve customer loyalty—Prevent the clients from switching to other competitors once MNP goes live.
- Increase/keep market share—Aggressively pursue customers of other companies enticing them to switch to the company's mobile plans.
- Improve quality of service—While the organization was a leader in the quality of service in the market, some of their equipment has aged and started to fail from time to time.
- Develop regions—The capital of the country was bringing the major flow of revenues to all the mobile players in the market. However, the management felt that considerable sales can be generated in the regions that remained underdeveloped.
- Public image—Maintain the company's image through social responsibility projects and sponsorship of youth programs.
- Increase revenue and profits—Was initially considered as an important strategic ingredient but was dropped from the list because it was included as one of the direct selection criteria (see the following text).

Scoring Model

The scoring model developed at the meeting consisted of the following variables (see also Table 8.1):

- Financial return on investment (ROI)
- Competitive advantage
- Improves customer satisfaction
- Innovativeness

Table 8.1 Eastern European Mobile Provider A Portfolio Scoring Matrix

Selection Criteria	Points Awarded			
	61 Points			
Joker	1 Point	5 Points	10 Points	Kill?
Financial (ROI)	ROI < 10%	10% ≤ ROI ≤ 20%	ROI > 20%	No
Competitive advantage	All competitors in the market are already offering this service	One or two competitors in the market are already offering this service	No other competitor is offering this service	No
Improves customer satisfaction	Does not affect customer satisfaction	Somewhat improves customer satisfaction	Significantly improves customer satisfaction	No
Innovativeness	The product or services has been on the international telecom market for more than 3 years	The product or services has been on the international telecom market for 1–2 years	The product or services has been on the international telecom market for less than 1 year	No
Strategic fit	Fits at least one of the strategic criteria	Fits two or three strategic criteria	Fits four or five strategic criteria	Yes
Time to market	More than 2 years	Between 1 and 2 years	Less than 1 year	No

- Strategic fit
- Time to market

The first criterion selected was the financial impact of the project (return on investment). This was one of the key variables for the strategy selection, but as was mentioned earlier, the sheer importance of this variable merited its inclusion in the scoring model itself. The managers agreed that the projects with the ROI less than 10% shall receive 1 point; projects with the ROI between 10% and 20%, 5 points; and finally, the ones with the ROI higher than 20%, 10 points. Interestingly enough, a very heated discussion occurred regarding the financial factor. A considerable group of senior managers felt that the projects with an ROI less than 0% should be killed automatically. However, mentioning of a simple fact that all of the maintenance projects, including upgrades of servers and other equipment, will be permanently removed from their project lists forced them to change their mind on the subject.

Competitive advantage was the next criterion included in the model. The proposals that involved a creation of the product or service already offered by all the company's competitors would receive 1 point, and if it was offered by one or two of the competing organizations, it would get a score of 5 points. And finally, if the proposed product or service was unique to the market, it would be awarded 10 points.

Since customer satisfaction played a major role in the company's strategy, this variable has also been added to the scoring model. The points were distributed in the following fashion:

■ Does not affect customer satisfaction—1 point
■ Somewhat improves customer satisfaction—5 points
■ Significantly improves customer satisfaction—10 points

Innovativeness of the product or service has also been deemed a very important factor in the project selection process. If the proposed product or service has been on the international telecom market for more than three years, it would get only one point in the current scoring model; if it has been around for one to two years, 5 points; and if it has been offered for less than one year, 10 points.

The executives have also decided that the strategic fit variable should also be added to the model to better align the projects with the overall company strategy. The projects that supported at least one of the strategic criteria received 1 point; those supporting between two and three criteria, 5 points; and the ones supporting four or all of them, 10 points. It was decided that this category should be a "kill" so that if the project proposal fits none of the objectives, it would be automatically dropped from the list.

Finally, considering the speed at which the telecommunications industry develops and changes in the modern times, the management felt a need to include a "time-to-market" variable to reward faster projects. The endeavors requiring more than two years of development received 1 point under the current system; the ones requiring between one and two years, 5 points; and the ones requiring less than a year, 10 points.

As a result, the maximum number of points that the proposal can receive in the scoring model is 60 points and the lowest is 6 points. The senior management team has also decided to adopt the "joker project" system where a proposal that scored low in the current matrix, but had a great breakthrough potential, would get an automatic score of 61 points, thus taking this project to the very top of the list. In addition, government-mandated regulatory projects would also fall into that category.

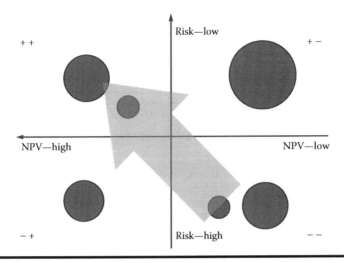

Figure 8.1   Eastern European mobile provider A portfolio balance—NPV vs. risk.

Portfolio Balance

Considering the fact that the risk factor scored fairly high when selecting the variables for the scoring model, but still did not make the cut, the executive team decided to include risk into the portfolio balance bubble chart (see Figure 8.1).

Strategic Alignment

The management team settled on a "top-down, bottom-up" approach for the strategic alignment of the company's projects. The strategic buckets designated by the executives were

- Breakthrough strategic projects—20%
- New products—10%
- Product improvements, extensions, and enhancements—20%
- Maintenance and fixes—30%
- Cost reductions—20%

Eastern European Mobile Provider B

The company discussed in this case study was at the time a fairly young organization with a very successful track record. In a short period of its existence, it managed to acquire approximately 20% of the local market and was competing very successfully with several other larger competitors.

Having said that, they have been experiencing several challenges that could probably be attributed to the rapid growth of the organization. One of the major issues facing the mobile provider was the low quality of the service and the resulting reluctance of potential new customers to use the company's services.

Another challenge was the stagnant revenue and profit growth. The board of directors have repeatedly mentioned this fact to the executive team and applied their pressure to improve the financial results.

An increase in the market share has also been viewed as one of the top priorities for the upcoming several years. One can probably see the connection between this strategic goal and the mission to improve the financial results mentioned in the previous paragraph.

Strategy

The company strategy for the upcoming three-year term consisted of the following goals:

- Increase market share—to boost the existing market share from 20% to around 30%–35%.
- Increase revenue and profits—as was mentioned earlier, the organization has been mandated to boost up their financial results.
- Improve the quality of services, products, and network—one of the major goals of the mobile provider to improve the quality of its services and its reputation among customers.
- Improve reputation—encourage social responsibility projects.
- Increase corporate base—obtain more of lucrative business customers.

Scoring Model

The scoring model developed by the executive team consisted of the following variables (see also Table 8.2):

- Strategic fit
- Financial
- Technical feasibility
- Market attractiveness
- Resources

One of the most important criteria mentioned by the management team was the strategic fit of the project proposal. This was done to ensure the proper alignment of all future endeavors with the company strategy for aggressive growth. It was decided that a project that incorporates one of the strategic goals would get 1 point in the current system. Projects fitting two or three of the criteria would be awarded 5 points and the ones including between four and five of the goals would be awarded 10 points. This category has also been designated as a "kill" criterion;

Table 8.2 Eastern European Mobile Provider B Portfolio Scoring Matrix

Selection Criteria	Points Awarded			
	51 Points			
Joker	*1 Point*	*5 Points*	*10 Points*	*Kill?*
Strategic fit	Fits at least one of the strategic criteria	Fits two or three strategic criteria	Fits four or five strategic criteria	Yes
Financial	Low	Medium	High	No
Technical feasibility	Difficult A lot of external expertise is required	Medium difficulty Some external expertise is required	Easy Only internal expertise is required	No
Market attractiveness	Low Very few requests from customers	Medium Average number of requests from customers	Strong Multiple requests from customers	No
Resources	70+ man-months	10–69 man-months	Less than 10 man-months	No

in other words, the projects that did not include any of the strategic goals were automatically removed from the list without considering other parameters.

The next category included was the financial factor (i.e., the ROI of the project). Despite the fact that this criterion has already been included in the strategic fit list, the executives insisted on adding this variable as an independent criterion to the scoring model basing this decision on the importance of the financial performance of the company.

The team could not come up with measurable criteria for this variable, so it was decided that the ranking points will be awarded in the following manner:

■ Low ROI—1 point
■ Medium ROI—5 points
■ High ROI—10 points

Considering the fact that the company has experienced certain issues with the reliability of their wireless network and IT systems, the management decided to focus more on the projects that did not require excessive external participation, motivating this decision that new and unknown technologies are inherently riskier than familiar ones. The proposal would receive 1 point for the project requiring a lot of eternal participation, 5 points for requiring some external expertise, and 10 points for a completely in-house project.

Market attractiveness was the next variable added to the model due to a demand to create attractive offerings for customers to increase the customer base—both through the market share and corporate accounts—and to improve the financial results. The points were distributed in the following fashion:

■ Low market attractiveness (i.e., very few requests from customers)—1 point
■ Medium market attractiveness (i.e., average number of requests from customers)—5 points
■ Strong market attractiveness (i.e., multiple requests from customers)—10 points

Finally, the management included the "resources" category to the scoring criteria to promote the so-called quick-win projects that would potentially generate additional income for the organization. According to their decision, the projects requiring an investment larger than 70 man-months would be awarded 1 point; those with human resource requirements in between 10 and 69 man-months, 5 points; and those requiring less than 10 man-months, 10 points.

The team has also agreed to implement the "joker" version of the projects for either the ideas that scored low in the matrix but were deemed to be strategic initiatives that had to be undertaken by the organization or the regulatory projects imposed by the local government.

Project Analysis

We actually had a chance to take several of the company's endeavors (current and forthcoming) and assess them using the new scoring model. The three projects considered were (see Table 8.3) as follows:

1. 4G—A transformation of the company to the new 4G (LTE) standard
2. Network upgrade—A major initiative to upgrade the entire wireless network infrastructure
3. Underground connectivity—Providing the company's customers to access voice and data services while riding on the local subway system

The first project considered was the 4G transformation already undertaken by the company. The project scored 10 out of 10 points in the strategic fit category, because according to the company's executives, it was expected to boost the organization's market share, increase its profits and revenues, improve the quality of services offered, and enhance the company's reputation. There also has been a heated discussion as to whether the LTE would assist the company with acquiring additional business customers. However, since 10 points were awarded for achieving at least four out of five strategic criteria, this discussion has been deemed irrelevant.

Table 8.3 Eastern European Mobile Provider B Project Analysis

	4G	*Network Upgrade*	*Underground*
Strategic fit	10	10	5
Financial	10	1	5
Technical feasibility	1	1	5
Market attractiveness	5	1	5
Resources	1	1	5
Total	**27/50**	**14/50**	**25/50**

Since the mobile company expected a major increase in revenues and profits from the LTE implementation, it was deemed that the project should be awarded 10 out of possible 10 points in the "financial" category.

Furthermore, the project scored only 1 out of 10 points in the "technical feasibility" category since the executives knew that the project would imply involving a significant external expertise.

In the "market attractiveness" category, the project has been awarded only 5 points since the senior management could not agree on how to count customer requests for the 4G transformation. On the one hand, they could not identify specific customer requests for the LTE, and yet on the other hand, they felt that many of their customers wanted higher download speeds and faster connections. Hence, it was decided to award the project only 5 out of the possible 10 points.

Finally, the category that did not generate any major discussions was the "resources" variable; all of the executives present agreed that the project would require way more than 70 man-months to accomplish.

Thus, the 4G (LTE) implementation project generated 27 out of the possible 50 points.

The next project assessed according to the newly developed portfolio model was the "network upgrade" initiative to address the connectivity and the quality of service issues mentioned earlier in this chapter.

While it managed to receive 10 out of 10 points in the "strategic fit" category—for its perceived impact on the market share, quality of services, reputation, and the corporate base—it came close to failure in all the remaining classes.

The project received 1 out of 10 in all the remaining categories because the expected ROI was negative (financial criteria), it required a lot of external expertise (technical feasibility), customers did not ask for it specifically (market attractiveness), and finally, it required a lot of human investment (resources).

As a result, the project was able to generate only 14 out of the possible 50 points, thus making it a very unlikely candidate for implementation. However, the management of the company including the CEO felt that it was one of the most important endeavors to be implemented by the organization. Therefore, the CEO, with the unilateral support of all the executives, invoked his "joker" power to award 51 points to this project, thus taking it to the very top of the project list.

The final project considered was the implementation of the underground network that would allow the subway riders to talk and download data on their mobile devices while using the subway system.

The executives decided that the project should receive 5 points for the "strategic fit" category since it only assisted in improving the quality of service and the reputation of the company.

There was a very heated discussion regarding the financial impact of this endeavor, but eventually the executives agreed that the best they could expect is a medium increase in profitability and to award it 5 points.

The project has also received 5 points in the "technical feasibility" category due to the need for some external expertise, 5 points for "market attractiveness," and 5 points for "resources" as it required approximately 40–50 man-months to accomplish.

Thus, the total value of the project was 25 out of possible 50 points. As a result of this exercise, the projects analyzed ranked in the following manner:

- Network upgrade—51 points (Joker)
- 4G—27 points
- Underground—25 points

Portfolio Balance

The executive board of the organization has decided to focus on the following factors when monitoring the project portfolio balance (see Figures 8.2 and 8.3):

- Strategic fit vs. net present value (NPV)
- Projects risk vs. NPV

Strategic Alignment

The management team settled on a "top-down, bottom-up" approach for the strategic alignment of the company's projects. The strategic buckets designated by the executives were

- Maintenance projects—20%
- Enhancement projects—50%
- New products/strategic projects—30%

Note: Regulatory projects imposed by the government would receive an automatic score of 51 points and together with other "joker" initiatives would not be included in any of the buckets.

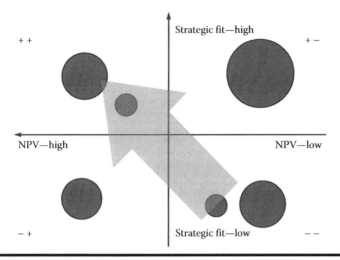

Figure 8.2   Eastern European mobile provider B portfolio balance—strategic fit vs. NPV.

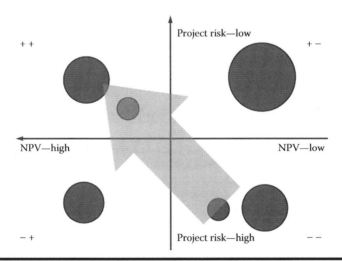

Figure 8.3   Eastern European mobile provider B portfolio balance—projects risk vs. NPV.

Central European Mobile Provider C

The next company to be discussed in this chapter is a telecom provider in the central European country. The country's telecom market is very competitive with between 40 and 60 million subscribers and with at least six companies operating in the market.

The organization was a recognized telecom service leader who claimed to have at least 30% of the market. The core business of the organization included fixed-network telephony, broadband Internet, mobile voice and data, and Internet Protocol television products and services. Such customer base and a wide array of services allowed the company to generate close to €2 billion in sales in 2012.

Furthermore, the company had a long tradition of being an innovation leader in the local market and could boast such new products as mobile payments and mobile cloud computing, to name a few.

The country's telecommunications market has also undergone some serious changes in the few years prior to holding the portfolio management workshop with the company's executives.

First, the country has experienced a continued growth in the availability and use of the Internet through not only computer but also increasingly popular mobile devices, such as smartphones. Second, after many years of dominance, the fixed-line phones started to lag behind the usage of mobile phones. And finally, the market has experienced an explosion in the growth of the cable and Internet protocol television.

Strategy

Based on the market position and the current situation in the industry, the executives felt that the following components should be incorporated into the organizational strategy:

- Continuing support of core businesses—To carry on maintaining the same levels of revenue
- New services introduction—To continue the aggressive penetration of the market and to expand their products and services portfolio (some of the directions mentioned were banking and cloud computing, but unfortunately the rest of the focus areas were kept confidential)
- Improving the brand awareness—A very important step considering multiple competitors in the market
- Generating savings of 10% in the operating expenses—To improve the financial results
- Improving the network quality—To continue to deliver exceptional services to customers and keep or increase the market share

Scoring Model

The scoring model developed by the company's senior management included the following variables (see Table 8.4):

- Financial
- Leverage of core competencies
- Innovativeness

Table 8.4   Central European Mobile Provider C Portfolio Scoring Matrix

Selection Criteria	Points Awarded			
	19 Points			
Joker	*1 Point*	*2 Points*	*3 Points*	*Kill?*
Financial	NPV < $5 million	$5 < NPV < $20 million	NPV > $20 million	No
Leverage of core competencies	Completely external resources	Some internal and some external resources	Internal resources only	No
Innovativeness	Addresses zero or four features requested by customers	Addresses five or seven features requested by customers	Addresses eight or more features requested by customers	No
Simple IT architecture	Very complicated IT infrastructure is required	Somewhat complicated IT infrastructure is required	Simple IT infrastructure is required	No
New segments?	Targets zero or one new segments	Targets two new segments	Targets three new segments	No
Availability of critical resources	Low	Medium	High	Yes

- Simple IT architecture
- New segments?
- Availability of critical resources

Despite the recommendations of the facilitator to assign scores that stood apart from each other considerably (e.g., 1, 5, and 15), the committee insisted on awarding 1, 2, and 3 points in each of the categories.

The financial factor was by far the most important variable in the model with almost all senior managers casting at least one of their votes for it. After much deliberation, the managers decided on the following ranges for the NPVs:

- The project with an NPV of less than $5 million would get a score of 1 point.
- The proposal with an NPV of less than 20 but more than $5 million would receive 2 points.
- The project with an NPV higher than $20 million receives 3 points.

The financial category was kept as a "no kill" due to the fact that the company must take on from time to time large projects (mainly in the technology field) with a negative NPV, also known as the "stay-in-business" projects.

Leverage of core competencies was another factor added to the mix with 1 point awarded to projects requiring a lot of external help, 2 points to the endeavors needing a few external resources, and 3 points to the projects requiring only internal resources. The executives have also decided not to designate this variable as a "kill" category for reasons similar to the one mentioned earlier.

Another interesting and fairly unique variable added to the matrix was the innovativeness factor. The executives claimed that they have a secret list of more than 50 features (i.e., new products or services) that have at 1 point of time been requested by their customers but had not been implemented yet. So, this category was designed to capture the innovativeness merit of the new projects in the following fashion:

- Project scope contains between zero and four of the new features—1 point
- Project scope contains between five and seven new features—2 points
- Project scope contains eight and more features requested by customers—3 points

Once more, the management decided not to mark this category as "kill" primarily due to the fact that large infrastructure upgrade projects may not cover any of the new features at all. However, the CIO of the company remarked that they might make the creation of the new features possible.

Since IT plays a major role in the operations of any telecom company, the management deemed necessary to include the simplicity (or complexity) of the required IT infrastructure as one of the factors. Due to the fact that it is fairly difficult to quantify the technology component of any given project, the decision was made to move ahead with the following qualitative criteria:

- Very complicated IT infrastructure is required—1 point
- Somewhat complicated IT infrastructure is required—2 points
- Simple IT infrastructure is required—3 points

IT infrastructure complexity was not designated as a "kill" category for the reasons mentioned earlier.

Another factor that could be viewed as a "strategic innovativeness" variable was added to the mix to reward the projects that addressed the three new strategic directions that the company decided to embark on in the near future. As was mentioned earlier, these included banking and possibly cloud computing, but not all of them have been disclosed during the session. The points have been assigned in the following manner:

- Targets zero or one new segments—1 point
- Targets two new segments—2 points
- Targets three new segments—3 points

Finally, the committee added, in their own opinion, one of the most important factors to the scoring model, the availability of critical resources. The executives have explained that for the past several years, the company has been facing a recurring problem: time and time again they approved multiple "attractive" projects only to find out that all of them shared key resources, primarily senior technical experts from the engineering and IT departments. Because of that, all of the projects in the pipeline suffered, people were forced to work long hours, and the product quality was low.

Therefore, the availability of critical resource variable has been included into the model with the following scores allocated to it:

- Low—1 point (i.e., the proposed project requires a lot of effort by the key technical resources)
- Medium—2 points (i.e., the proposed project requires several key technical resources)
- High—3 points (i.e., none or very few of the key technical resources are required)

This was the only criteria designated as a "kill" category by the company management in the newly developed scoring model. This implied that if several key resources supposed to be engaged on a proposed venture were unavailable, the whole initiative would potentially be scrapped altogether. Having said that, it was noted that if the project scored very high in all other categories, the portfolio team would consider dropping or delaying other projects to put the one question forward.

As in many previous cases described in this book, the executive team reserved the right to designate an occasional "joker" project, an endeavor that scored relatively low in the current model, but the senior management felt that the project had a potential to become a strategic breakthrough for the organization.

To sum up all the aforementioned points, a project candidate at the company could generate as much as 18 points and as little as 6 points in the current scoring model. Another final comment to make about this particular scoring model is that it did not include the "strategic fit" category into the matrix. However, one can argue that the variables selected in one way or another aligned with the company's strategy. For example, support of core businesses and 10% of savings in OPEX have been addressed by the financial variable, while the new service initiative has been taken care by the innovativeness and new segment variables.

Project Analysis

Again, in this particular case, the executive committee and the facilitator had the luxury of taking several of the company's projects currently being considered for implementation and analyzing them using the newly developed scoring model. The three projects selected for the assessment were

Table 8.5 Central European Mobile Provider C Project Analysis

	Mobile Wallet	*CRM*	*New Tariff Plan*
Financial	3	1	3
Leverage of core competencies	2	1	2
Innovativeness	3	2	3
Simple IT architecture	1	2	3
New segments?	3	2	1
Availability of critical resources	3	2	1
Total	**15/18**	**10/18**	**13/18**

- Mobile wallet—An application that would enable customers to pay for goods and services using their mobile phones
- CRM—Customer relationship management system implementation
- New tariff plan—A new tariff plan to be developed for a specific customer segment

The first project considered was the "mobile wallet" initiative (see Table 8.5) that was expected to enable the company's customers to pay for various goods and services using their mobile devices. The executives expected this project to generate an NPV of more than $20 million; hence, the initiative has been awarded 3 out of possible 3 points in the "financial" category.

The management was expecting to handle most of the project using internal resources with some outside help; thus it was given 3 out of 3 points for the "leverage of core competencies."

According to the managers, this initiative would address more than 15 new features requested at various points of time by customers. Therefore, the project received 3 points in the "innovativeness" category.

Furthermore, the senior management was convinced that the IT infrastructure that needed to be created for this project would be fairly simple. Interestingly enough, the facilitator of the portfolio management session had a slightly different opinion on the subject having been through similar projects before. As a result of this argument, the team has invited the leading systems architect who happened to be familiar with the project and who confirmed that this initiative would require a major upgrade to the existing IT infrastructure as well as a significant development effort. Therefore, after a long deliberation, the portfolio committee decided to award this project only 1 out of 3 points for the "IT architecture" category.

Also, according to the top management, the project proposed was expected to target at least three of the new segments identified by the company. As a result, the project received 3 points in the "new segments" category.

Finally, all the critical resources have been freely available for this project and the initiative received 3 points in this category as well.

To sum up, this project has gathered 15 out of the possible 18 points, thus making it a "star" endeavor in the company portfolio.

The next project considered was the "CRM implementation" initiative lobbied by both the sales and the marketing departments of the company.

This endeavor received only 1 point for the "financial" factor, since the executives could not be convinced by the sales and marketing people that the project would generate a significant NPV for the organization.

A major external involvement was required on this project since the company was considering buying the CRM solution from a vendor who was expected to deploy a team of its professionals to configure and fine-tune the system. Therefore, the project—yet again—received only one point in this category.

As far as the "innovativeness" factor was concerned, the initiative received 2 points from the team due to the fact that it addressed at least five of the features requested by customers.

The IT infrastructure component was deemed to be of significant enough size to warrant a score of 2 points.

The project proposal has also received 2 points each in the "new segments" and "critical resource availability" categories. The management felt that they would be able to target two new market segments, while the key resources would be partially available for the project.

Thus, this endeavor gathered 10 out of possible 18 points in the newly developed scoring model, making it, at least temporarily, an outsider of the current portfolio.

The final project considered was the new tariff plan that was supposed to target a new large customer segment.

With respect to the financial aspect, the executives were convinced that the new plan was capable of generating an NPV in excess of $20 million and, despite numerous challenges by the facilitator who did not feel this estimate was realistic, continued insisting on their estimate. Finally, after about half an hour of deliberations, the team decided to proceed with a mark of 3 out of possible 3 points for this variable.

The management felt that some eternal expertise would be required on this endeavor, thus awarding this project a score of 2 points for the "leverage of core competencies" category.

Since the project was expected to include at least 20 new features requested by customers, it easily received 3 points in the next category—the "innovativeness" factor.

Also, since the project did not require any significant changes to the existing IT infrastructure, it has easily received 3 points for that category.

Furthermore, the project was not expected to target any of the new segments, so it has only been awarded 1 point in the "new segments" category.

In the "availability of critical resources" category, the proposal was awarded 1 point as well since it required several key company resources that were not readily available.

To sum up, the project collected 13 out of the possible 18 points. This result has placed the initiative somewhere close to the mid to upper levels of the organizational portfolio.

Portfolio Balance

Interestingly enough, just like in the previous case study, the company management unanimously agreed that the following dimensions should be considered when assessing the project portfolio balance (see Figures 8.4 and 8.5):

- Strategic fit vs. NPV
- Risk vs. reward (NPV)

Strategic Alignment

The mobile provider's managers chose to utilize the very popular top-down, bottom-up model for the portfolio alignment. However, their approach to the bucket distribution was somewhat unique; rather than designating one silo for all of their new product families (breakthrough) projects, they decided to allocate a separate bucket for each of the family. As a result, their project distribution looked like as follows:

- Mobile finance
- Cloud computing

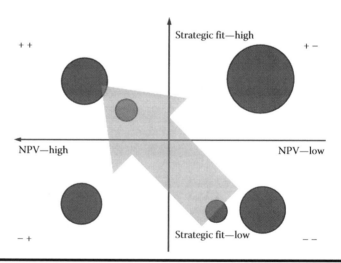

Figure 8.4 Central European mobile provider C portfolio balance—strategic fit vs. NPV.

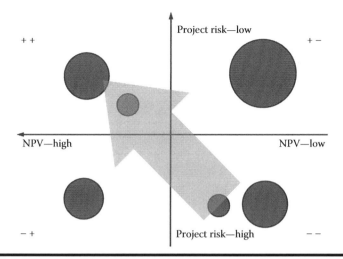

Figure 8.5 Central European mobile provider C portfolio balance—risk vs. reward (NPV).

- Other new product families
- Technical debt/improvements in the existing products
- Maintenance and upgrades

Western European Mobile Provider D

The fourth company to be discussed in this chapter is a mobile provider operating in a relatively small European country. The mobile company is a provider of a range of fixed-line, mobile, data, and Internet communication services in a highly competitive market.

Over the course of the previous three or so years, the organization has experienced a decline in its overall revenues (more than 6% on an annual basis). However, most of this decline can be attributed to an array of external rather than internal factors. First, the market that the company was operating was one of the most competitive in the world with more than 5 mobile operators and more than 20 Internet providers, all in the market with less than 15 million subscribers.

Second, the country's government has exerted a lot of pressure on the mobile providers by passing legislation especially on mobile termination rates and roaming tariffs. The combination of these two factors led to a drop in the average revenues generated per customer by more than 10%.

Despite the external challenges and a significant drop in revenues, the organization was able to turn around its fortunes due to cost reductions in operations, targeting new segments and bundling various products and services to fit various target markets.

Strategy

Considering their previous success, the executives of the company have decided to continue focusing on the three strategic pillars that had helped them to turn around their fortunes in the previous years. These included the following:

■ Continuing cost reductions in core business areas—To address the problem of declining revenues and to shift capital to the development of new products and services
■ Increase revenues from converged infrastructure—By packaging multiple information technology components into a single, optimized computing solution both to further reduce operating costs and to improve IT flexibility for future product and service development
■ Active sales channels into new segments—To increase revenue generating

Scoring Model

The scoring model developed by the company executives was quite interesting as it contained only three variables (see also Table 8.6):

1. Strategic fit
2. Complexity (How many external resources will be required?)
3. Economic impact (EI) (expected annual revenue)

The first category included was the strategic fit that consisted of three main initiatives (see earlier texts). The portfolio committee has decided that the proposals

Table 8.6 Western European Mobile Provider D Portfolio Scoring Matrix

Selection Criteria	Points Awarded			
	31 Points			
Joker	1 Point	5 Points	10 Points	Kill?
Strategic fit	Zero strategy components	One or two strategy components	Three strategy components	No
Complexity (How many external resources will be required?)	>6	3–5	0–2	No
Economic impact	EI < €0.5 million/year	€0.5 < EI < €1.9 million/year	EI > €2.0 million/year	No

including none of the strategic initiatives would get 1 point in the new portfolio scoring model; the ones with either 1 or 2 strategy components, 5 points, and finally the ones with all three strategic components, 10 points. The executives were reluctant to mark this variable as a "kill" category and decided to wait for a year and to run several dozen projects through the model to get a feel as to whether this category should become a "kill."

Next, the management included the complexity variable to the model. As in many portfolio scoring cases, they have decided to measure the complexity of the future project by assessing the number of the external resources that would need to be involved in it. They felt that this approach would serve two purposes at once. First, it would decrease the risks associated with any given endeavor, and second, it would eliminate or at least penalize costlier projects. The points have been allocated in the following fashion based primarily on the historical data from the past projects:

- Projects requiring more than 6 external resources—1 point
- Projects requiring more between 3 and 5 external resources—5 points
- Projects requiring 2 or less external resources—10 points

The complexity variable has been marked as neither a "kill" category nor certain "cost of doing business" and technology projects that the company management was planning in the near future needed a massive vendor participation.

Finally, the third ingredient in the scoring mix was the so-called economic impact (EI), which basically stood for incremental revenue expected to be generated by the proposed product or service. After much deliberation, the committee resolved to go ahead with the following point allocation:

- EI < €0.5 million/year—1 point
- €0.5 < EI < €1.9 million/year—5 points
- EI > €2.0 million/year—10 points

Again due to the fact that the executives expected to undertake several technology upgrade projects that were not expected to generate any tangible revenue, this variable has not been designated as a "kill."

The executive committee reserved the right to grant a "go ahead" for the projects that scored low on the scoring matrix, but there was a consensus that they could be the next breakthrough endeavors that could take the company to the "next level."

Therefore, the maximum score that the project proposal could get under this model is 3 points, while the maximum is 30. However, if the executive "joker" is used, the proposal automatically rises to the top of the list with a score of 31 points.

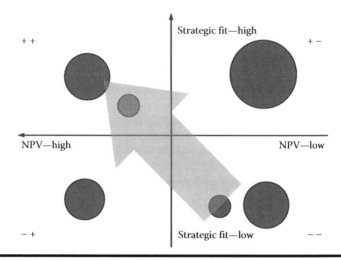

Figure 8.6 Western European mobile provider D portfolio balance—strategic fit vs. NPV.

Portfolio Balance

The executive board of the organization has decided to focus on the following factors when monitoring the project portfolio balance (see Figure 8.6):

■ Strategic fit vs. NPV

Strategic Alignment

The management team settled on a "top-down, bottom-up" approach for the strategic alignment of the company's projects. The strategic buckets designated by the executives were

■ Market development—40%
■ Operational excellence—10%
■ Sustainability (maintenance)—5%
■ Transformation—45%

Summary

At the beginning of this chapter, we discussed the overall global state of the telecommunications industry and the outlook for the next few years. Based on the predictions of the professional experts, the telecommunications companies must pay a special attention to the following aspects of their businesses to survive and thrive in the ever-changing mobile business:

- Migrate from 3G to 4G and improve the quality of the connection.
- Create new and innovative products and services.
- Implement aggressive marketing of their new product and services.
- Develop competitive pricing and improve customer service.

The first company we analyzed, having already completed a transformation from 3G to 4G, had to concentrate on the remaining initiatives. The quality of the service factor has been included in the strategic fit list. The creation of the new products and services has been included in the scoring matrix under the innovativeness category. The company has also covered the aggressive marketing component by including initiatives like "improving customer loyalty," "increasing the market share," and "developing regions" in the company's strategy.

Finally, one can argue that the organization has not directly addressed the competitive pricing issue by mentioning it neither in the strategies list nor in the scoring model itself.

The second organization we examined has listed the 4G project as one of its key priorities in the upcoming year, thus addressing the 3G to 4G transformation issue. By the same token, the "improvement of the connection quality" requirement has been addressed by the "improvement of the quality of services, products, and network" ingredient in the strategy.

It can be argued that social responsibility projects and the program to obtain more of the lucrative business customers can be construed as the organization's response to the "aggressive marketing" key to success.

Furthermore, the need to develop new and innovative products and services has been directly addressed in the portfolio scoring mode by including the "market attractiveness" threshold to evaluate new project ideas.

Unfortunately, just like in the previous case study, the executives of the company failed to include a strategy component or a scoring variable that would have been responsible for the "competitive pricing" requirement.

The third organization considered in this chapter has also completed the 4G migration project by the time of the portfolio session. By the same token, the inclusion of the "continuing support of core businesses" initiative has addressed the requirement for the quality of the connection.

The innovation requirement mentioned at the beginning of this chapter has been enthusiastically embraced by the company's management; it resulted in the addition of a strategy component called "new services introduction" as well as inclusion of not one but two ingredients in the scoring model: innovativeness and new segments variables.

Improving the brand awareness strategic initiative was a response to the need to implement aggressive marketing.

Interestingly enough, none of the executives had deemed necessary to include aggressive pricing as either the strategy component or scoring model variable.

Finally, let us take a look at the fourth organization in this chapter and attempt to assess its response to the ever-changing situation in the telecom industry.

The company has also completed the 4G migration by the time the portfolio workshop took place, hence addressing the first requirement on the list. The organization at the time had an excellent track record with respect to service quality levels; therefore, it is also understandable why the management decided not to focus on this aspect.

The organization has also indirectly encouraged investment into new products and services by trying to reduce costs in the core business areas to shift freed-up capital into the development. One can also argue that the strategic initiative aimed at converged infrastructure would have also enabled more flexibility with innovation.

The marketing requirements have been addressed by the inclusion of the strategic component targeting sale channels into new segments.

Again, as in the previous three case studies, the management failed to include the competitive pricing factor into either the strategies list or the scoring model.

Chapter 9

Project Portfolio Management in the Government and Not-for-Profit Sector

Government and Not-for-Profit Sector Overview

It is estimated that the combined spending of all the global government hovers around $20–$25 trillion. Considering the financial and political challenges that our world has encountered in the last 10 years, one must ask the key question: How is the taxpayers' money being spent to provide all the stakeholders with the best potential mix of products and services?

According to one of the leading consulting companies, there are several areas where both the government agencies and not-for-profit agencies must focus their efforts to navigate the dangerous waters of the modern life (KPMG International, 2011).

Fiscal responsibility—Governments of all levels must use all of their efforts to balance their budgets and reduce deficit spending by managing their budgets. In addition, they need to identify new sources of revenues and execute their long-term strategies to capture and sustain them.

Cost efficiency—An item closely related to the fiscal responsibility, mentioned earlier. While, on the one hand, public sector organizations will have to develop and nurture new sources of revenues, on the other hand they will be forced to decrease their costs and/or improve their operational efficiencies.

Maintenance and development of infrastructure—A major challenge for the developing countries that have gone years and sometimes decades without sufficient investment into the infrastructure. It looks like investment into building and upgrading new roads, bridges, tunnels, water supply, sewers, electrical grids, telecommunications, and so forth will become one of the major items on the government's "to do" list.

New technologies—Another domain closely related to the fiscal responsibility is the ability of public sector organizations to embrace the new technologies, including mobile, e-government, and Internet. Governments must re-examine the way they are organized and redefine their operating models to create more efficient structures that will better anticipate and meet the needs of their constituents, stakeholders, and business communities.

Transparency and accountability—Yet another sphere that could be linked to both the cost efficiency and fields of new technologies. Due to the direct link between good governance and efficient performance, all of the stakeholders, including both the lawmakers and the taxpayers, are now demanding more accountability and transparency from the federal, state, provincial, municipal, and other levels of the government.

Risks—The governments worldwide must actively manage the risks of their projects. These include not only the performance-related risks (e.g., budget, time, and scope) but also risks related to the reputational, operational, legal, and regulatory domains.

Sustainable environment—With the apparent climate change issues on hand, all of the government and other nonprofit organization must actively seek new ways of conserving energy, development of carbon-neutral services, and strategic investments into alternative energy developments.

Government and Not-for-Profit Sector Case Studies

Introduction

In this chapter, we will examine the strategies and the portfolio models of four organizations from around the world. The list of agencies includes the financial department of the ministry of defense, European mortgage and lending agency, IT department of the Canadian university, and one of the European central banks.

Ministry of Defense: Financial Department

The organization described in this section is an accounting and financial services agency reporting directly to a ministry of defense of one of the countries. It was conceived as a financial services provider or the civil and military members of the ministry.

The agency has recently experienced a major shift in strategy with the government insisting on increasing the revenues, cutting the costs, and improving the operational efficiency. It has also committed to improving the ratio of their successful projects. In addition, improvement of customer satisfaction has also been deemed one of the key priorities for the organization.

Strategy

Based on the facts mentioned earlier, the organization's strategy has been defined by the executive managers for the next five years and consisted of the following components:

- Be ready for the systems audit in 2017
- Increase organizational customer base
- Increase and diversify sources of revenues
- Decrease operational costs and increase efficiency
- Improve customer satisfaction
- Attract and develop the best professionals

Scoring Model

The scoring model development exercise with the organization's executives yielded the following variables:

- Strategic fit
- Resources required
- Financial value
- In-house knowledge/project complexity
- Risks

As a result of the calibration exercise, we developed the following scoring matrix (see Table 9.1).

Strategic fit has been the variable that received the most votes during the model development exercise. The executives decided to award 1 point to the project proposal that fits one or two of the strategic initiatives, 5 points for the projects covering three to four of the initiatives, and 10 points to the endeavors that covered five or more of the organizational strategies. In addition, this category was deemed to be a "kill" type, which meant that unless the project was defined as a regulatory (i.e., mandatory) or a joker initiative, it would get an automatic kill if it fits no organizational strategies at all.

Table 9.1 Ministry of Defense Portfolio Scoring Matrix

Selection Criteria	Points Awarded			
	51 Points			
Joker	1 Point	5 Points	10 Points	Kill?
Strategic fit	Fits one or two of the strategies	Fits three or four of the strategies	Fits five or more strategies	Yes, unless regulatory or "joker" project
Resources required	X > 300 man-months	200 < X < 299 man-months	X < 200 man-months	No
Financial value	IRR < 5%	5 < IRR < 7.5%	IRR > 7.5%	No
In-house knowledge/ project complexity	Very complex project Requires a lot of external knowledge	Medium complexity Some outsourcing is required	Simple project No or very little external expertise is required	No
Risks Reputational Operational Financial Legal	Very risky project that carries significant reputational, operational, and financial risks	Somewhat risky project that carries at least two of the risks	Low-risk project, may carry one of the three risks	Yes, if legal risks are present

The organization decided to include the "resources required" variable as well to penalize larger initiatives requiring a lot of human resources (HRs). As a result, it was decided that the projects needing more than 300 man-months of internal effort would receive only 1 point in the scoring model, while the projects requiring between 200 and 300 man-months would be awarded 5 points. Finally, the projects where the effort required was less than 200 man-months would receive full 10 points.

The next variable to be included in the model was the "financial value" of the proposed project. Despite the nature of their work and the fact that the majority of their projects did not have a positive net present value (NPV), the executives felt that adding this variable would directly reflect their agency's effort to cut costs and increase revenues as has been outlined in the strategic plan. The points have been distributed in the following manner:

- Internal rate of return (IRR) < 5%—1 point
- 5 < IRR < 7.5%—5 points
- IRR > 7.5%—10 points

The in-house knowledge/project complexity factor has also been added to the mix since the senior management felt that their organization had been taking on a lot of large and very complex projects that used up almost all of the agency's resources.

If the proposed initiative was deemed to be very complex and requiring a lot of external knowledge, it would get a score of 1 point. If the project was of medium complexity and some outsourcing was needed, it would get a score of 5 points. However, if the project was simple and no or very little of external expertise was needed, it would get the full 10 points.

Finally, due to the sensitive nature of their work, the executives felt that they needed to consider a range of risks in their project filtration mechanism. As a result of a very long discussion, they agreed to include the following risks to the model:

- Reputational
- Operational
- Financial
- Legal

The points have been distributed in the following manner:

- Very risky project that carries significant reputational, operational, and financial risks—1 point
- Somewhat risky project that carries at least two of the risks—5 points
- Low-risk project that may carry one of the three risks—10 points

As a result, the minimum number of points the project proposal can receive under this scheme was 5 points and the maximum possible was 50 points. In addition, the management decided to include the "joker" project concept into the model. A score of 51 points would be awarded to the regulatory and legal initiatives or the projects that fall into "the next big breakthrough" or "cost of staying in business" categories.

Portfolio Balance

Considering the complexity of their operations as well as the number and the size of their projects, the senior management decided to monitor the following dimensions of the portfolio balance (see Figures 9.1 through 9.3):

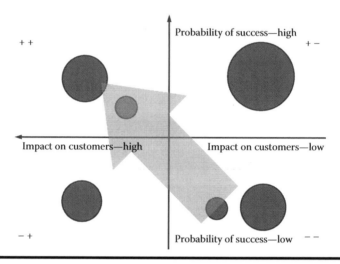

Figure 9.1 Ministry of defense portfolio balance—probability of success vs. impact on the customers.

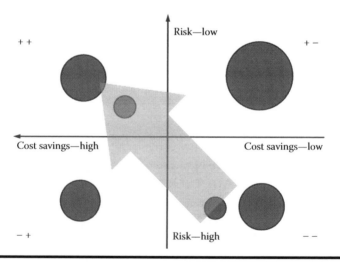

Figure 9.2 Ministry of defense portfolio balance—risk vs. cost savings.

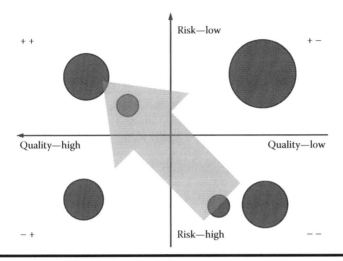

Figure 9.3 Ministry of defense portfolio balance—risk vs. quality.

- Probability of success vs. impact on the customers
- Risk vs. cost savings
- Risk vs. quality

Strategic Alignment

Like many other companies described in this book, the organization decided to adopt the "top-down, bottom-up" approach and designate the following portfolio buckets:

- Regulatory projects—As needed
- Other projects (with the resources remaining after fulfilling the regulatory projects' needs)
    - Military pay projects—50%
    - Maintenance projects—20%
    - System modernization projects—20%
    - Strategic breakthrough projects—10%

Note: All percentages allow for a ±5% variation.

Federal Loan and Mortgage Lending Agency: Eastern Europe

The second organization to be discussed in this chapter is the mortgage and housing agency located in one of the eastern European countries. The main objective of the company is to implement governmental programs for providing affordable housing and ensuring improvement of housing conditions for the population.

In addition, the organization was under considerable pressure to increase its revenue while reducing the costs. Furthermore, the company was expected to improve its service offerings and decrease the operational risks.

Strategy

The organization's strategy has been very well defined by the executive board and consisted of the following elements:

- Increase the segment of families that could buy their own house/condo to 65% by 2016.
- Decrease the spread between the mortgage-backed security rates and final borrower rates to 2.2% by 2016.
- Increase the percentage of mortgages that are funded by the mortgage-backed securities to 55% by 2016.
- Increase the return on equity (ROE) to 4.5% by 2016.
- Cut the cost-to-income ratio to 20% by 2016.

Scoring Model

The executives of the agency decided to include the following variables into the scoring model during the project portfolio management workshop:

- Project size
  - Budget
  - HRs
- Strategy fit
- Financial (IRR)
- Social value
- Risks

The first variable added to the model was the size of the proposed project (see Table 9.2). The senior management felt that the organization has been chronically taking on large and very complex projects in the past years and saw the need to skew the portfolio balance toward small- and medium-sized endeavors. As a result, the point distribution looked as follows:

- Projects with a budget exceeding $10 million and/or HR investment of more than 50 man-months—1 point
- Projects with a budget between $2 and $120 million and/or HR investment of between 10 and 50 man-months—5 points
- Projects with a budget less than $2 million and/or HR investment of less than 10 man-months—10 points

Table 9.2 Federal Loan and Mortgage Lending Agency Portfolio Scoring Matrix

Selection Criteria	Points Awarded			
	51 Points			
Joker	*1 Point*	*5 Points*	*10 Points*	*Kill?*
Project size Budget Human resources	B > $10 million HR > 50 man-months	$2 million < B < $10 million 10 < HR < 50 man-months	B < $2 million HR < 10 man-months	No
Strategy fit	Fits at least one of the strategies	Fits two or three of the strategies	Fits four or five of the strategies	Yes
Financial (IRR)	IRR < 10%	10 < IRR < 20%	IRR > 20%	No
Social value	Does not include any of the government programs	Includes at least one government program	Includes two or more government programs	No
Risks	High	Medium	Low	Yes, if budget >$10 million and high risk

The second variable added to the model was the strategic fit of the project. It has been decided that the proposal that fits one of the strategic initiatives would receive 1 point; a proposal that covers two to three strategies, 5 points; and the endeavor including between four and five of the strategic initiatives, full 10 points. In addition, this criterion has been designated as a "kill" category; in other words, the proposals covering none of the strategic initiatives, unless they fall into the regulatory or "joker" categories, would be automatically removed from further consideration.

Due to the government pressure to become more fiscally responsible and look for creative ways to increase the revenue and decrease the costs, the management decided to add a financial factor to the model in the form of IRR. The points would be assigned in the following manner:

- IRR < 10%—1 point
- 10 < IRR < 20%—5 points
- IRR > 20%—10 points

The fourth variable included, the social value factor, has been added for two reasons. First, the executives felt that it would improve the overall image of the organization, and second, they thought it might have a direct impact on the customer satisfaction levels. Therefore, the senior management of the agency decided that the project proposal supporting a number of the government-sponsored social value programs would receive a score of 1 point, while a project including one social responsibility initiative would receive 5 points, and the one that supports two or more of the social programs would get a score of 10.

Finally, the managers made a decision to incorporate the project risk factor into the selection model. The projects that were deemed to be of high risk would be awarded 1 point; projects that carried a medium risk, 5 points; and the projects with low risk, 10 points. In addition, it was decided that a combination of "high risk" in the risk category and an expected budget of more than $10 million in the "project size" category would place the proposal into the "kill" category, as the organization wanted to get rid of large, complex, and high-risk projects.

Project Analysis

To "test" the newly developed scoring model, we decided to run several projects through it and assess the official results by comparing them to the management's "gut feel." The three projects selected were

1. Modular housing project—An initiative that involved buying the new fast building technology and deploying it in the country to provide people with cheap and quality housing
2. Enterprise resources planning (ERP) system project—A deployment of the new ERP system
3. Mortgages for teachers project—Involved a creation of a new financial product for the country's school teachers that would enable them with easier access to mortgages

The "modular housing" project received 1 point out of possible 10 in the size category due to the fact that its budget was expected to by far exceed the $50 million threshold (see Table 9.3).

In the strategy category, it received 5 out of 10 possible points since it was forecasted to incorporate the following strategic initiatives:

Table 9.3 Federal Loan and Mortgage Lending Agency Project Analysis

	Modular Housing	*ERP System*	*Mortgages for Teachers*
Project size	1	5	1
Strategy fit	5	5	1
Financial (IRR)	10	1	1
Social value	10	1	5
Risks	1	1	10
Total	27/50	13/50	18/50

- Increase the segment of families that could buy their own house/condo to 65% by 2016.
- Increase the ROE to 4.5% by 2016.
- Cut the cost-to-income ratio to 20% by 2016.

Since it was deemed to include more than two of the social responsibility programs, the project proposal received 10 out of potential 10 points in the social value category.

Finally, considering the project size and complexity, it got only 1 point in the risk category for a total value of 27 out of possible 50 points.

The ERP project was much easier to score since the scope and the budget of the initiative have already been known in detail. The following is the layout of the scores it received in each one of the categories:

- Project size—1 point, since it was expected to cost more than $10 million
- Strategy fit—5 points, because the management expected it to improve the overall efficiency and improve both the ROE and cost-to-income figures
- Financial (IRR)—1 point, because it is practically impossible to create a meaningful financial model for such a project
- Social value—1 point, because it did not include any of the social initiatives
- Risks—1 point, because it was expected to be a very risky and complex project
- Total—13 points out of 50

The mortgages for teachers project received the following scores:

- Project size—1 point, since it was expected to cost less than $50 million
- Strategy fit—1 point, because it was expected to support the "increase the segment of families that could buy their own house/condo to 65% by 2016" strategy
- Financial (IRR)—1 point, because it was not expected to generate any meaningful revenues for the agency

- Social value—5 points, because it included a social initiative by the government to assist the low-income groups of population
- Risks—10 points, because it was a low-risk project
- Total—18 out of 50 points

Portfolio Balance

The executives decided that they would like to monitor the portfolio using the following dimensions (see Figures 9.4 and 9.5):

- Strategic fit vs. IRR
- Risks vs. IRR

Strategic Alignment

The organization's senior management resolved to utilize the "top-down, bottom-up" methodology and designated the following buckets for the new project proposals:

- Regulatory and "joker" projects—as needed
- Other projects (with the resources remaining after fulfilling the regulatory projects' needs)
- Maintenance projects—50%
- Strategic improvement projects—50%

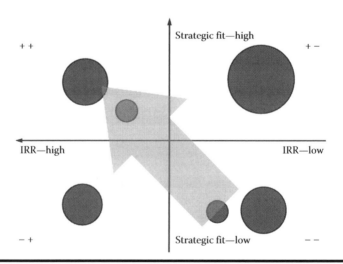

Figure 9.4 Federal loan and mortgage lending agency portfolio balance—strategic fit vs. IRR.

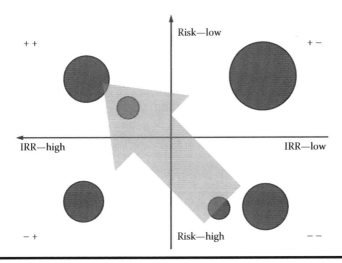

Figure 9.5   Federal loan and mortgage lending agency portfolio balance—risks vs. IRR.

Canadian University: IT Department

The third organization to be featured in this chapter is an IT department of a relatively small Canadian university. The university in general and the IT department in particular have been experiencing several challenges over the course of several years before the portfolio management initiative has been implemented at the organization.

The university's executive management reported very stiff competition for prospective students from both Canadian and U.S. colleges located nearby. In addition, there was a general feeling at the top level that the value of the projects they delivered was low, although they did not have any tools to objectively assess the worth of their initiatives. Furthermore, the university's IT department has been overloaded with projects originating from the numerous departments, schools, and offices at the organization while their resource pool has been fixed.

Strategy

During the first part of the workshop, it was determined that the university has a mission consisting of several goals. These included the following:

■ To attract more Canadian and international students to the university
■ To improve the university's reputation in Canada and internationally
■ To provide the best possible mix of services and benefits to the students
■ To provide the best possible mix of services and benefits to the employees
■ To increase social value of programs and initiatives undertaken at the university

Scoring Model

Table 9.4 contains the key criteria (i.e., variables) that will be used by the university's IT department to score and rank all the project proposals generated by both internal (i.e., IT) and external (i.e., non-IT) entities at the university. All of the criteria have been generated during the portfolio management working session.

There are five regular criteria that include the following:

1. Strategic fit of the proposed project
2. Resources required for the proposed project
3. Technical feasibility of the proposed project
4. Financial value of the proposed project
5. Riskiness of the proposed project

If the project scores very positively in any particular category, it receives a score of 15 points for that particular criterion (e.g., great strategic fit or low resource requirements).

If the project receives a negative rating in any particular category, it gets a score of 1 point (e.g., very difficult to implement from technology standpoint or a very high risk rating).

If the project receives an average or neutral rating in a particular category, it gets a score of 5 points (e.g., a project with a mid-level NPV). Therefore, the highest possible score a project proposal can get is 105 points (5 criteria × 15 points per criteria) and the lowest possible score is 5 points (5 criteria × 1 points per criteria).

In addition to the five regular criteria, the workshop attendees found it necessary to include one special "super criterion" to the existing mix of variables to account for "must do" and/or regulatory projects that have to be undertaken by the university to ensure business continuity. This criterion would award 106 points to the project proposal and would automatically promote it to the top of the list of the proposed endeavors.

Project Analysis

We were able to conduct an analysis of the largest ongoing project at the university at the time. The project in question was the university ERP system upgrade. The ERP project received the following points:

■ Strategic fit—15 points (great strategic fit)
■ Resources required—1 point (very high resource requirements)
■ Technical feasibility—1 point (very complex endeavor)
■ Financial value—5 points (probably mid-level cost savings over a lifetime of the system)
■ Riskiness —1 point (very risky project)
■ Total—23 out of possible 105 points

Table 9.4   Canadian University Portfolio Scoring Matrix

Selection Criteria	Points Awarded		
Must do or regulatory A strategic "must do" initiative that has to be implemented irrelevant of all other selection criteria. Must be approved by the executive-level management	106 points		
	1 Point	*5 Points*	*15 Points*
Strategic fit Does the proposed project fit one or more of the criteria below: Attracts students Improves the university's reputation Provides benefits to students Provides benefit to employees Increases social value?	Low Fits zero or one of the criteria	Medium Fits two or three of the criteria	High Fits four or more of the criteria
Resources required What level of investment would the proposed project require in terms of human effort and/or financial costs?	High More than 300 man-hours and/ or more than $15,000 cost	Medium Between 200–300 man-hours and/or $10,000– $15,000 cost	Low Between 100–200 man-hours and/or $5,000– $10,000 cost
Technical feasibility How familiar are the university's IT employees with the technologies inherent in the proposed project?	Very difficult A significant external expertise will be required. The technology is completely or almost unknown to the university's IT employees	Somewhat difficult Will need external expertise, not all technologies involved are familiar to the university's IT employees	Easy Can be implemented by internal IT employees, known technology

(Continued)

Table 9.4 (*Continued*)   Canadian University Portfolio Scoring Matrix

Selection Criteria	Points Awarded		
	1 Point	*5 Points*	*15 Points*
Financial value What are the financial benefits of the proposed project? Will it increase revenues and/or cut costs and by how much?	Minor Revenue generation/cost (including total cost of ownership [TCO]) saving initiative with a 0 < NPV < $100,000	Medium Revenue generation/ cost (including TCO) saving initiative with a $100,000 < NPV < $1,000,000	Major Revenue generation/ cost (including TCO) saving initiative with an NPV > $1,000,000
Riskiness How risky is the proposed project? Does it contain any reputation, regulatory, financial, or operational disruption risks?	High Significant reputation, regulatory, financial, or operational disruption risks	Medium Some reputation, regulatory, financial, or operational disruption risks	Low Little or no reputation, regulatory, financial, or operational disruption risks

However, it was understood by everyone that this was a "must do" project for the university and had to be undertaken irrelevant of its scores in all other categories, simply due to the fact that the university would have ceased to operate due to the outdatedness of the old system.

Portfolio Balance

The project balance component of the university IT portfolio management framework will probably be the least restrictive of all dimensions (i.e., project value, portfolio balance, and strategic alignment) and will be reported to the senior management to assess the general utilization of IT resources with respect to their impact on the university's business.

The potential dimensions the university executives wanted to monitor were (see Figure 9.6):

- Risk vs. reward
- Academic vs. research vs. system support projects

Strategic Alignment

It was decided by the management team that the balance of the university's IT project portfolio should be assessed based on the magnitude of their impact on the university's business. These categories include the following:

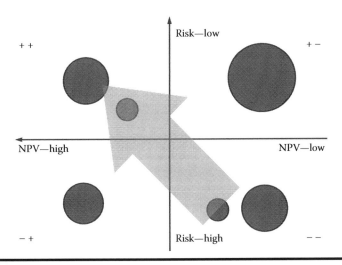

Figure 9.6   Canadian university portfolio balance—NPV vs. risk.

- Maintenance projects (30%)—The cost of doing business projects that may include, but are not limited to, hardware or network upgrades, replacement of obsolete equipment, and regular maintenance of the existing systems.
- Growth projects (50%)—Projects that improve the university's operations and/or business at a tactical level. These projects may include, but are not limited to, a program to attract students from a specific country, construction of a small building on campus, or an improvement to the existing system.
- Transformation projects (30%)—Projects that have a strategic impact on the university's operations and business. These projects may include, but are not limited to, a deployment of university-wide ERP system (i.e., banner), opening of a new faculty or school, construction of a major building, opening of a new campus, or a major overhaul of an existing system.

The "top-down, bottom-up" methodology was chosen by the senior management for the strategic alignment process.

Company D: European National Bank

The organization discussed in this section is the central bank of one of the European countries and a part of the European System of Central Banks.

The main function has thus become banking and financial supervision. The objective is to ensure the stability and efficiency of the system and compliance to rules and regulations; the bank pursues it through secondary legislation, controls, and cooperation with governmental authorities.

Other functions include market supervision, oversight of the payment system and provision of settlement services, the treasury service, economic analysis, and institutional consultancy.

Several issues have been dogging the organization for several years preceding the portfolio management initiative. These included

- Euro-integration—Due to the integration of all European banks into the European System of Central Banks, the agency was expected to fulfill a long list of obligations including fine-tuning the coordination and cooperation mechanisms; adapting the rules, procedures, and processes for supervision; and strengthening specialist skills.
- Economic stability—Maintain the economic stability in the country in times of economic troubles in other European countries by employing appropriate fiscal and monetary policies.
- Fiscal responsibility—Decrease costs incurred by the bank by improving operational efficiency.
- Low quality of services—Address the ongoing complaints from the local private sector banks regarding the quality of service offered by the central institution.

Strategy

As a result of these challenges, the central bank came up with the following four-pronged strategy:

1. Continue strengthening the bank's role within the Eurosystem.
2. Improve services for the wider community (includes improving procedures for consumer protection, increasing preventative measures and initiatives to increase the quality of the services offered by payment service providers).
3. Review costs, rules, and procedures to increase efficiency and decrease operational costs.
4. Adopt diversity as a corporate value and diversity in terms of gender, age, ability, and expertise/experience.

Scoring Model

The scoring model developed by the executive team during the portfolio management workshop consisted of five variables (see Table 9.5):

- Eurosystem policy alignment
- National policy alignment

Table 9.5   European National Bank Portfolio Scoring Matrix

Selection Criteria	Points Awarded			
	51 Points			
Joker	1 Point	5 Points	10 Points	Kill?
Eurosystem policy alignment How well is the proposed project aligned with the Eurosystem policies and guidelines?	Weak Covers at least one of the Eurosystem policies or guidelines	Medium Covers two of the Eurosystem policies or guidelines	Strong Covers three or four of the Eurosystem policies or guidelines	No
National policy alignment How well is the proposed project aligned with the national policies and guidelines?	Weak Covers at zero or one of the national policies or guidelines	Medium Covers two or three of the national policies or guidelines	Strong Covers four or more of the national policies or guidelines	No
Strategic fit Does the proposed project fit one or more of the criteria below: Strengthen the bank's role within the Eurosystem Improve services for the wider community Increase efficiency Diversity	Low Fits one of the criteria	Medium Fits two or three of the criteria	High Fits four of the criteria	Yes, unless regulatory or "joker" project
Resources How many resources (both $ and HR) will the project require?	Large B > €500,000 HR > 100 man-months	Medium €250,000 < B < €500,000 50 < HR < 100 man-months	Small B < €250,000 HR < 50 man-months	No

(Continued)

Table 9.5 (Continued) European National Bank Portfolio Scoring Matrix

Selection Criteria	Points Awarded			
	51 Points			
Joker	1 Point	5 Points	10 Points	Kill?
Risks Does the implementation of the proposed project carry the following risks: Operational Reputational Financial Legal/compliance	High The project carries at least three of the risks	Medium The project carries one or two of the risks	Low The project carries no risks	Yes If medium to high legal/ compliance risk

- Strategic fit
- Resources
- Risks
  - Operational
  - Reputational
  - Financial
  - Legal/compliance

The first variable, "Eurosystem policy alignment," was added to speed up the integration into the overall European banking system of the organization. The idea was that the more Eurosystem policies and guidelines the project proposal covers, the more points it receives.

The second variable was expected to address the issue of aligning the bank's projects with the internal national policies. If the project covered between zero and one of the national policies, it would get a score of 1 point; if it included two to three policies, a score of 5 points; and if four or more of the policies, a score of 10 points.

The strategic fit variable was designed to analyze the alignment of the project with the internal strategies. The point breakdown was as follows:

- Fits one of the criteria—1 point
- Fits two or three of the criteria—5 points
- Fits four of the criteria—10 points

This variable has been designated a "kill" category for the projects including none of the bank's strategies, unless this was deemed to be a regulatory or a "joker" project.

The fourth criterion selected was the project size. As in many previous cases, the executives felt that their organization should focus on smaller, less complicated endeavors and move away from the long, costly internal "megaprojects." The project size points have been distributed in the following manner:

- Budget > €500,000 and HR > 100 man-months—1 point
- €250,000 < budget <€500,000 and 50 < HR < 100 man-months—5 points
- Budget < €250,000 and HR < 50 man-months—10 points

Finally, the executives decided to add the risk variable to the scoring mix to weed out riskier, less predictable initiatives.

Portfolio Balance

The bank executives decided to monitor the portfolio balance using the following dimensions (see Figures 9.7 through 9.9):

- Financial risk vs. reward
- Operational risk vs. reward
- Reputational risk vs. reward

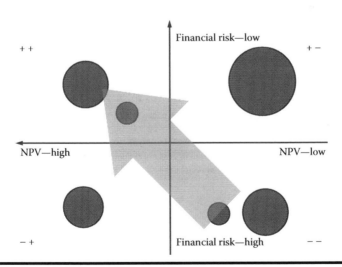

Figure 9.7 European national bank portfolio balance—financial risk vs. reward.

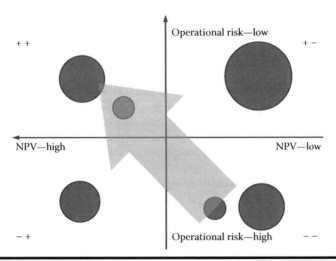

Figure 9.8 European national bank portfolio balance—operational risk vs. reward.

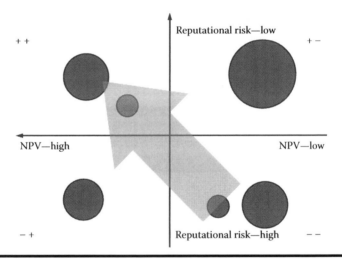

Figure 9.9 European national bank portfolio balance—reputational risk vs. reward.

Strategic Alignment

The executive board decreed that they would like to utilize the following buckets for the portfolio strategic alignment:

- Policy alignment and regulatory projects—50%–60%
- Maintenance projects—15%–20%
- New initiative projects—20%–35%

Summary

We mentioned at the beginning of the chapter the key areas in which the government and not-for-profit sectors should be investing in the next several years to remain competitive and provide their customers or taxpayers with the best possible mix of products and services. These include

- Fiscal responsibility
- Cost efficiency
- Maintenance and development of infrastructure
- New technologies
- Transparency and accountability
- Sustainable environment

Let us now review the four case studies we have examined in this chapter and analyze whether their strategies and portfolio choices reflected the aforementioned list.

The ministry of defense strategies focused on increase and diversification of the sources of revenues, decrease of the operational costs, efficiency increase, and improvement of customer satisfaction. Furthermore, their scoring model included variables like resources required (i.e., focus on smaller, more efficient initiatives), financial value, and risks.

The European mortgage and lending agency's strategic focus has been improving the service for its customers (increase the segment of families that could buy their own house/condo to 65% by 2016) and on their financial performance (increase the ROE to 4.5% by 2016 and cut the cost-to-income ratio to 20% by 2016).

In addition, the agency's scoring variables included such factors as risk, financial performance of the project, and social value of their initiatives.

The Canadian university's strategy included improving the overall quality of education and services and a strong focus on the social value of its programs.

Also, the university's scoring model included variables like resources required, financial value, and riskiness of the proposed projects.

Finally, the European central bank included the following components into its strategy:

- Service improvements
- Increasing efficiency and decreasing operational costs

In addition, the organizational scoring model incorporated risk and resource factors to decrease the size of their projects and improve their risk management.

Chapter 10

Project Portfolio Management in the Professional Services Industry

Professional Services Sector Overview

Professional services account for a large proportion of GDP, particularly in developed economies. Professional, scientific, and technical services accounted for 7.5% of GDP in the United States, for example, in 2010, while the real estate sector accounted for a further 12.7% (Qfinance, 2013).

Worldwide IT spending is projected to total US$3.7 trillion in 2013, a 4.2% increase from 2012 spending of US$3.6 trillion, according to a forecast published by Gartner, Inc. in 2013 (Gartner, 2013).

According to the industry experts, the organizations operating in the professional services sector should be aware regarding several key trends in the environment (Dawson, 2005).

Client sophistication—The clients are expected to get more aware about the services offered to them. This implies several outcomes. First, this implies that rampant cross- and up-selling of your services to "fleece" your customers is less likely to work. It also means that, assuming your company is offering quality services, the sales process should get easier, since the client knows exactly what he wants and hopefully appreciates the value of the services provided. Another implication stemming from this trend is that the sophisticated clients will demand a wider array of services fine-tuned specifically to their needs.

Governance—After the Enron, WorldCom, Arthur Andersen, and Parmalat debacles, the governments around the world, especially those in developed countries, are expected to stiffen the laws and regulations regarding organizational governance and transparency. As a result, the organizations in the professional services sector should be prepared to accept and deliver a large number of regulatory projects.

Connectivity—Rapid development of communication technologies had a profound effect on the business world in the last several years. Again, these developments have, at least, a dual impact on the companies around the world. First, customers would expect faster services, and second, the organizations would need to consider serious investments into their information technology infrastructure.

Modularization—Another trend that is expected to happen in markets is called "unbundling of the services." This implies that customers rather than buying all-in-one solutions would tend to break down their needs and outsource them to different vendors. For example, a company that used to outsource all of its legal needs to one organization may now separate them into two packages: simple tasks and complex tasks. The simpler activities could be outsourced to a smaller firm that is willing to charge a lower price. Again, it should force the established professional services businesses to create new products and services to close the potential gap caused by the unbundling.

Globalization—With the great advances in the communication technologies, it has become extremely easy to shop for professional services beyond the city and even the country the organization operates in. The lesson learned from this trend is that if you do not provide a quality service for competitive price, someone in China, India, or eastern Europe will. Once more the answer to this challenge is to create unique services that cannot be easily copied by the vendors abroad and to cut costs.

Professional Services Case Studies

Introduction

In this chapter, we will focus on project portfolio management models developed by three professional services organizations: a professional services department of a European software product development company, a European IT services organization, and an IT department of a global consulting firm.

Professional Services Company A: European Software Company

The first company to be discussed in this chapter is the European software company that produced several microcredit applications targeted mainly at the telecom and banking sectors. The product development division of the firm was responsible for the development of several microcredit products, while the professional services team has been charged with the deployment and the fine-tuning of their platform on the client company sites.

The organization has been quite successful for several years with its products in the market, but rising competition and decreasing service fees forced the company management to reassess their approach to the selection and prioritization of their professional services engagements.

Strategy

The strategy for the professional services division developed by the senior management was very simple and straightforward and presented in the form of three questions:

1. Will this project open new geographic markets for our company?
2. Will this project provide us with access to the new industries?
3. Will this project enable us to cross-sell our other products to the client?

Scoring Model

The portfolio scoring model developed during the project portfolio management workshop included the following variables (see Table 10.1):

- Strategic importance
- Financial forecast
- Risks
- Internal influence

The first variable added to the model was the strategic fit of the proposed project. The points have been distributed in a straightforward fashion: if the project fits one of the company's strategies, it would get 1 point in the scoring model; if it fits two of the strategies, it would get 5 points; and, finally, if it covered all three of the strategic initiatives, it would get all 10 points. This variable was designated as a "kill" category. All projects that covered none of the company's strategic initiatives had to be approved by the senior management.

The second variable added was the expected return on investment (ROI). The points have been distributed in the following manner:

- 10% < ROI (internal rate of return [IRR]) < 15%—1 point
- 15% < ROI (IRR) < 20%—5 points
- ROI (IRR) > 20%—10 points

This variable has also been designated a "kill." All projects with an ROI less than 10% could not go ahead unless preapproved by the company's executives.

The next category in the mix was the project risks. The management came up with a list of six most important risks and distributed the matrix points depending

Table 10.1   European Software Company Portfolio Scoring Matrix

Selection Criteria	Points Awarded			
	41 Points			
Joker	1 Point	5 Points	10 Points	Kill?
Strategic importance How important is the project for the overall strategy of the company? Will it introduce us to new geographic markets and industries or allow cross-sell opportunities?	Low Fits one of the strategies	Medium Fits two of the strategies	High Fits all three of the strategies	Yes, unless a "joker" project
Financial forecast What are the financial benefits[a] of the proposed project? Will it increase profits and by how much?	Minor 10% < ROI (IRR) < 15%[b]	Medium 15% < ROI (IRR) < 20%	Major ROI (IRR) > 20%	Yes, unless a "joker" project
Risks How many of the following risks is the project exposed to provide 1. Potential management change at the client company 2. High probability of the political upheaval in the country 3. Potential or existing regulatory restrictions 4. Lack of IP protection or high competition 5. Nonpayment risk	Major four[c]	Medium two or three	Moderate zero or one	Yes, unless a "joker" project
Internal influence How significant is our company's influence at the client company?	Low Low or nonexistent level of influence	Medium Moderate level of influence	High Significant influence	No

[a] Also consider the absolute value of the expected revenue.
[b] An ROI of less than 10% requires a special executive management approval.
[c] Number of risks exceeding 4 requires a special executive management approval.

on how many risks the project has been exposed to. The points were allocated in the following fashion:

- Four risks—1 point
- Two or three risks—5 points
- Zero or one risks—10 points

Once more, this was also a "kill" category for the projects receiving more than four risks. A continuation of this project required special approval from the company's senior managers.

Final, and somewhat unusual, variable added to the model was designated the strength of a relationship the software producer had with the management of the prospective client. If the influence was low or nonexistent, the proposal would get 1 point in the model; if there was a moderate level of influence present, the project would receive 5 points; and if there was a strong relationship between the senior managers of both companies, the project would be awarded full 10 points.

Portfolio Balance

The executive team has indicated that they would be interested in analyzing their portfolio from the following perspectives (see Figures 10.1 through 10.3):

- Financial value (ROI) vs. resources
- Financial value (ROI) vs. time to completion
- Strategic fit vs. risk

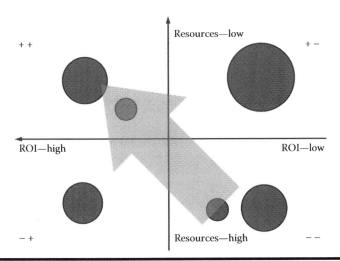

Figure 10.1 European software company portfolio balance—financial value (ROI) vs. resources.

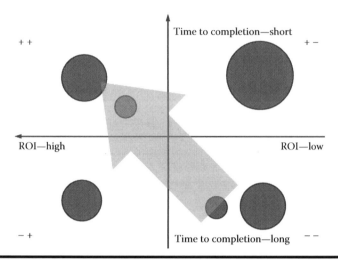

Figure 10.2 European software company portfolio balance—financial value (ROI) vs. time to completion.

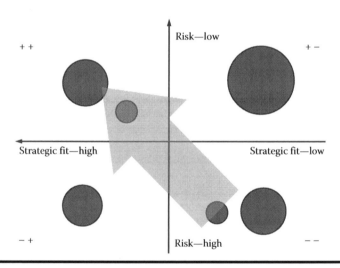

Figure 10.3 European software company portfolio balance—strategic fit vs. risk.

Strategic Alignment

The company executives decided that approximately 70% of the total technical team effort should be directed toward implementation (professional services) projects, while 30% of the total effort must be invested into the new products (product development).

Considering this breakdown and the number of technical resources in both the "new product" and implementation teams, the approximate resource allocation can be calculated in the following manner:

Total resources available = 8 men × 10 months* = 80 man-months

Resources available for project work = 80 man-months × 80%†
= 64 man-months

Resources available for implementation (professional services)
= 64 man-months × 70% ≈ 45 man-months

Professional Services Company B: European IT Services Company

The second professional services company to be discussed in this chapter is a fairly large European IT services provider. At the moment of the portfolio management initiative, the company has already been successfully operating in several countries and had a workforce of close to 500 employees.

The management felt that one of the main issues that project portfolio management can address is a large number of project requests coming from their customers that needed to be prioritized. They wanted to know which projects receive the highest priority, which ones can be postponed, and which engagements needed to be rejected.

Strategy

The strategy developed by the executive committee has been very simple and consisted only the following three points:

- Is the project request originating from the IT-dependent industries like the following?
  - Banking
  - Telecom
  - Government
  - Transportation
  - Insurance
- Will this project allow us to generate at least $100,000 in profits?
- Is this project coming from a company located in countries A, B, C, D, or E?

Scoring Model

The executives of the company produced the following project scoring model at the end of the facilitated portfolio management workshop (see Table 10.2):

- Strategic fit
- Technical feasibility

---

* To account for vacations, holidays, and sick days.
† It is assumed that approximately 20% of total technical team time is taken by the nonproject, business-as-usual work.

Table 10.2   European IT Services Company Portfolio Scoring Matrix

Selection Criteria	Points Awarded			
Must do or regulatory A strategic "must do" initiative that has to be implemented irrelevant of all other selection criteria and must be approved by the executive-level management	51 Points			
	1 Point	*5 Points*	*10 Points*	*Kill?*
Strategic fit Does the proposed project fit one or more of the strategic criteria?	Low Fits zero or one of the criteria	Medium Fits two or three of the criteria	High Fits all four of the criteria	No
Technical feasibility What is the complexity level of the project?	High Project involving application development	Medium Hardware and/ or off-the-shelf software	Low Hardware only project	No
Leverage of core competencies How familiar are our employees with the technologies inherent in the proposed project?	Low A significant external expertise will be required. The technology is completely or almost unknown to our employees	Normal This will need external expertise; not all technologies involved are familiar to our employees	High It can be implemented by our employees, which is considered as known technology	Yes, unless a regulatory or "joker" project

(Continued)

Table 10.2 (*Continued*)   European IT Services Company Portfolio Scoring Matrix

Selection Criteria	Points Awarded			
	1 Point	*5 Points*	*10 Points*	*Kill?*
ROI What are the financial benefits of the proposed project? Will it increase profits and by how much?	Minor 10% < ROI < 15%	Medium 15% < ROI < 20%	Major 20% < ROI < 25%	Yes, if ROI < 10% unless a regulatory or "joker" project
Probability of winning the bid What are the chances of this project being awarded to our company?	High There is a very low chance of project being awarded to our company (e.g., less than 30%)	Medium The chances of getting the project are fair (e.g., around 50%)	Low The chances of getting the project are high (e.g., more than 70%)	Yes, if the probability of winning is very low unless a "joker" project

- Leverage of core competencies
- ROI
- Probability of winning the bid

The first variable selected was the strategic fit. Since the company strategy has been very clear and well defined, the points allocation has been fairly straightforward:

- Fits one of the criteria—1 point
- Fits two or three of the criteria—5 points
- Fits all four of the criteria—10 points

Technical feasibility was the next variable added to the scoring model. The management felt that projects involving application development had a tendency to be overly complicated and the organization did not have sufficient internal capacity to handle them. Therefore, the projects that involved a significant portion of software development received 1 point; the projects including a mix of hardware installation and some off-the-shelf software installation and configuration, 5 points; and the initiatives involving only hardware installation, 10 points.

Leverage of core competencies was the next category included in the model. The executives felt that the sales managers were taking on too many projects where the technologies were completely unknown to the company employees, thus leading to a high percentage of troubled or failed projects. Therefore, the points in this category were awarded in the following manner:

- A significant external expertise will be required. The technology is completely or almost unknown to our employees—1 point.
- This will need external expertise; not all technologies involved are familiar to our employees—5 points.
- Can be implemented by our employees. The technology is known—10 points.

This category has been designated as a "kill" variable unless it was a regulatory or a "joker" project explicitly approved by the senior management.

The ROI was the next—very logical—variable added to the matrix since the company has been earning practically 100% of its revenues from their professional services projects. The points were awarded in the following fashion:

- 10% < ROI < 15%—1 point
- 15% < ROI < 20%—5 points
- 20% < ROI < 25%—10 points

This category has also been designated as a "kill" for the projects with ROI less than 10% unless they fell into the regulatory or a "joker" category.

Finally, due to the management's concern that the company was wasting a lot of its resources on bids for projects with very low probabilities of winning, they decided to add the fifth variable—probability of winning the bid. If the probability was less than 30%, the project would get 1 point. The projects that had a probability of winning close to 50% would get 5 points and the ones with a probability of winning are deemed to be more than 70%. This was also a kill category if the probability of winning was less than 10%.

Portfolio Balance

The management chose the risk vs. reward (ROI) diagram to monitor the balance of their professional services portfolio (see Figure 10.4).

Strategic Alignment

The company executives decided not to use any of the strategic buckets alignment models due to the following reasons: first, they felt that the strategic alignment has been covered during the scoring exercise, and second, they already had a designated "bucket" of professional services resources.

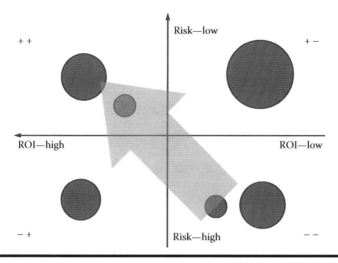

Figure 10.4   European IT services company portfolio balance—ROI vs. risk.

Professional Services Company C: IT Department of a Global Professional Services Company

The final organization to be discussed in this chapter is a global IT department of an international consulting company. The IT management's main goal was to prioritize their multiple projects and to properly align them with the company's strategy. The overall feeling was that the company had too many, frequently large and complex, internal technology initiatives on the go and needed to somehow pick the highest value ones to implement.

Strategy

As a result of the situation described earlier, the company's management came up with the following four-point strategy for the organization:

1. Improve profitable growth.
2. Maintain one-firm, cross-functional approach (cross-sell our services).
3. Improve operational efficiency (save money).
4. Motivate and retain the best employees.

Scoring Model

The executive committee decided to select the following five variables for the portfolio scoring model (see Table 10.3):

1. Strategic fit
2. Enables revenue growth

Table 10.3 Global Professional Services Company Portfolio Scoring Matrix

Selection Criteria	Points Awarded			
Must do or regulatory A strategic "must do" initiative that has to be implemented irrelevant of all other selection criteria and must be approved by the executive-level management	*51 Points*			
	1 Point	*5 Points*	*10 Points*	*Kill?*
Strategic fit	Fits one strategy	Fits two of the strategies	Fits three or more of the strategies	Yes, unless regulatory or "joker" project
Revenue growth enabled	Low The impact on revenue growth is nonexistent or negligible	Average Some positive impact on the revenue growth is expected	High A significant positive impact on the revenue growth is expected	No
Cost efficiency	Low or no impact	Average	High	No
Resources required	200+ man-days	61–200 man-days	<60 man-days	No
Time to market	4+ months	2–4 months	<2 months	No

3. Cost efficiency
4. Resources required
5. Time to market

The first variable included in the model was the strategic fit of the proposed project. The proposal would receive 1 point if it fits one of the strategies, 5 points if it covered two strategies, and full 10 points if it addressed three or more of the company's strategic initiatives. This category has been designated as a "kill" variable that would be waived only if it was a regulatory project or a "joker" initiative.

Also, the senior managers wanted to assess the project's impact on the revenue growth. The points were awarded in the following fashion:

- The impact on revenue growth is nonexistent or negligible—1 point.
- Some positive impact on the revenue growth is expected—5 points.
- A significant positive impact on the revenue growth is expected—10 points.

The next variable thrown into the mix was a potential project impact on the company's cost efficiency. The points in the model have been distributed in the following manner:

- Low or no impact—1 point
- Average—5 points
- High—10 points

Finally, another two variables added to the model to promote smaller, less complex projects were the resources required and the time to market. The points for these variables have been allocated as follows:

- Project effort > 200 man-days—1 point
- 61 < Project effort < 200 man-days—5 points
- Project effort < 60 man-days—10 points

and

- T > 4 months
- 2 < T < 4 months
- T < 2 months

Project Analysis

We actually had a chance to run a couple of the company's upcoming projects through the newly created scoring model. Here are the project candidates (see Table 10.4):

- New company website—An overhauled main website that had to conform to all search engine optimization standards. The main purpose of the site was to showcase multiple company services and promote them to potential customers.

Table 10.4   Global Professional Services Company Project Analysis

Categories	New Website	New Sales App
Strategic fit	10	1
Enables revenue growth	5	10
Cost efficiency	5	10
Resources required	1	5
Time to market	5	5
Total	26/50	31/50

■ New sales app—The new app was supposed to be installed on all the employee iPads. The idea was that whenever a company employee has a meeting with a client and learns that the customer company needs some additional services, she can quickly access all the relevant information via the sales app and show it to the client.

The website project received the following points in each of the categories:

■ Strategic fit—10 points because it, in the eyes of the executives, covered the "profitable growth," "cross-selling," and "improved operational efficiency" initiatives
■ Revenue growth enabled—5 points because it carried an average positive impact on the company's sales
■ Cost efficiency—5 points because the executives did not expect it to have a major impact on the cost savings
■ Resources required—1 point because it definitely required more than 200 man-days to accomplish
■ Time to market—5 points because the IT department estimated that the project would take between 3.5 and 4 months

The sales app project performed in the following manner:

■ Strategic fit—1 point because it only supported the "cross-selling" initiative
■ Enables revenue growth—10 points because the executives expected to project to have a major impact on sales
■ Cost efficiency—10 points because the executives—unlike the facilitator—continued to insist that the usage of the app would cut a lot of expenses
■ Resources required—5 points due to expected effort of 50–100 man-days

■ Time to market—5 points since the project was expected to last for about three months

Portfolio Balance

In order to scrutinize the portfolio balance, the executives decided to monitor the following bubble chart (see Figure 10.5):

■ Strategic fit vs. ROI

Strategic Alignment

The company decided to employ the following bucket breakdown for the strategic alignment purposes:

■ Regulatory projects—As needed
■ Joker projects—As needed
■ Maintenance projects—20% of the resources left after regulatory and joker initiatives
■ Operational efficiency projects—70% of the resources left after regulatory and joker initiatives
■ New products and services projects—10% of the resources left after regulatory and joker initiatives

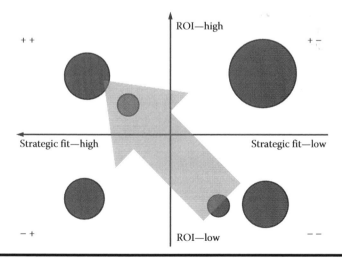

Figure 10.5 Global professional services company portfolio balance—strategic fit vs. return on investment.

Summary

The overview of the professional services industry at the beginning of this chapter demonstrated that the organizations operating in this industry would have to address the following demands in the foreseeable future:

- Customers will be expecting a wider array of services fine-tuned specifically to their needs.
- Customers will be demanding that their services are delivered faster and cheaper than before.
- The professional services companies must be able to create new products and services and move into new markets, by service, by industry, and by geography.
- Cost cutting in view of stiffening global competition, commoditization, and unbundling of services will be one of the major priorities.
- Due to the growing focus on governance, the companies will have to be prepared to accept and deliver a large number of regulatory projects.
- Investments into their information technology infrastructure are needed to better serve their customers.

Let us now examine the strategic and the portfolio scoring models created by the companies discussed earlier in this chapter.

The European software company added expansion into new geographic markets, new industries, and the ability to cross-sell the products as the key priorities in its strategy. Furthermore, inclusion of financial and risk variables was aimed to address the challenges of improving the profit margins and cutting the costs.

The second professional services organization analyzed in this chapter included both industry-based (banking, telecom, government, transportation, and insurance sectors) and geography-based (four specific countries) expansions into their strategy. In addition, they included several risk factors (technical feasibility, leverage of core competencies, and probability of winning the bid) into their scoring model, which aimed primarily to cut their losses and expenses on their external-facing projects. Finally, the ROI variable added to the portfolio model was aimed to improve the financial performance as well.

The third organization, the IT department of a global consulting firm, included profitable growth into the strategic priorities as well as improving operational efficiency to their scoring model to address shrinking revenues and cost-cutting concerns.

In addition, the company made a choice to improve its cross-selling capabilities to increase the revenue per client results.

Finally, time to market was yet another variable added into the scoring model to speed up the delivery of new products and services.

SUMMARIZING IT ALL

Chapter 11

Statistical Summary and Analysis

Introduction

This chapter summarizes and systemizes the results of the project portfolio management studies both at the industry and aggregate levels.

Pharmaceutical Industry

Scoring Models

The pharmaceutical industry was represented by three companies that demonstrated the following results:

Company A

- Strategic fit
- Market attractiveness
- Competitive advantage
- Technical feasibility
- Financial (sales)
- Risk
- Sales force readiness

Table 11.1   Pharmaceutical Industry: Most Popular Scoring Variables

Variable	Frequency (%)
Financial value (e.g., NPV, IRR, ROI, payback)	100
Strategic fit	67
Market attractiveness	67
Risk	67
Technical feasibility	67

Company B

- Market attractiveness (how many patients are out there?)
- Strategic fit
- Innovativeness
- Risk (both technical and market)
- Effectiveness
- Cannibalization
- Core competencies
- Competitors
- Financial (revenue)

Company C

- Innovativeness (financial benefits vs. risks)
- Candidate for China/Brazil/Russia?
- Resources

The most popular scoring model variables are presented in Table 11.1.

Portfolio Balance

A total of five models were employed by the three companies in the study.

Company A

- Market attractiveness vs. technical feasibility
- Return on investment (ROI) vs. probability of success

Company B

- Probability of success vs. total cost
- Probability of success vs. remaining cost

Table 11.2   Pharmaceutical Industry: Most Popular Balance Dimensions

Variable	Frequency (%)
Reward	80
Risk	40

Company C

- ROI vs. risks

The most popular portfolio balance variables are presented in Table 11.2.

Strategic Alignment

All three companies in the sample selected the "top-down, bottom-up" approach for the strategic alignment.

Product Development Industry

Scoring Models

The product development industry included seven companies with the following results:

Product company A

- Strategic fit
- Possible synergies
- Financial value
- Technical complexity
- Market attractiveness
- Competition and intellectual property

Product company B

- Strategic fit
- Leverage of core competencies
- Financial forecast
- Market attractiveness
- Synergy with other projects/products

Product company C

- Market attractiveness
- Fit to existing supply chain
- Product and competitive advantage
- Technical feasibility
- Time to break even
- Net present value (NPV)

Product company D

- Strategic fit
- Financials (NPV)
- Market attractiveness
- Technical feasibility

Product company E

- Strategic fit
- Time to market
- Market attractiveness
- Technical feasibility
- Competitive advantage
- Fit to existing supply chain

Product company F

- Strategic alignment
- Customer need
- Synergies with the existing business
- Technical feasibility
- Profitability (payback)
- Commercial/technical risk

Product company G

- Strategic alignment
- Financial benefit (NPV)
- Resource needs
- Risk if not executed

The most popular scoring model variables are presented in Table 11.3.

Table 11.3   Product Development Industry: Most Popular Scoring Variables

Variable	Frequency (%)
Strategic fit	86
Financial reward	86
Customer/market attractiveness	86
Synergies	71
Technical feasibility	71
Competitive advantage	43

Portfolio Balance

A total of seven models were employed by the seven companies in the study.

Product company A

■ Market or technical risk vs. reward

Product company B

■ ROI vs. resources

Product company C

■ Risk vs. reward

Product company D

■ NPV vs. cost

Product company E

■ Risk vs. reward

Product company F

■ Risk vs. reward

Product company G

■ Risk vs. reward

The most popular portfolio balance variables are presented in Table 11.4.

Table 11.4 Product Development Industry: Most Popular Balance Dimensions

Variable	Frequency (%)
Reward	100
Risk	71

Strategic Alignment

All seven companies selected the "top-down, bottom-up" approach for the strategic alignment.

Financial Industry

Scoring Models

The financial industry was represented by four companies with the following results:

Eastern European bank 1

- Strategic fit
- NPV
- Payback
- Execution risk

Western European bank

- NPV
- Payback
- Strategic fit
- Technical risk
- Customer impact (market attractiveness)
- Employee impact

North American brokerage company

- Strategic fit
- Revenue
- Time to market
- Project size and cost (resources)
- Existing expertise (leverage of core competencies)

Table 11.5   Financial Industry: Most Popular Scoring Variables

Variable	Frequency (%)
Strategic fit	100
Financial reward	100
Risk	60
Customer/market attractiveness	40
Technical feasibility	40

Eastern European bank 2

- Strategic fit
- ROI
- Market attractiveness/competitive advantage
- In-house expertise
- Risk and complexity
- Improves operational efficiency?

The most popular scoring model variables are presented in Table 11.5.

Portfolio Balance

A total of six models were employed by the four companies in the study.

Eastern European bank 1

- Payback vs. execution risk
- NPV vs. execution risk

Western European bank

- Technical risk vs. NPV

North American brokerage company

- Revenue vs. time to market
- Revenue vs. resources

Eastern European bank 2

- Risk vs. reward

The most popular portfolio balance variables are presented in Table 11.6.

Table 11.6 Financial Industry: Most Popular Balance Dimensions

Variable	Frequency (%)
Reward	100
Risk	67

Strategic Alignment

All four companies selected the "top-down, bottom-up" approach for the strategic alignment.

Energy and Logistics Industry

Scoring Models

Five companies were in the energy and logistics industry with the following results:

Energy company A

- Strategic fit
- Competitive advantage
- Market attractiveness
- Technical feasibility
- Financial (payback)

Energy company B

- Strategic fit
- Market attractiveness
- Competitive advantage
- Leverage core competencies
- NPV
- Payback
- Commercial risk

Energy company C

- Age of the technology platform or system
- Business implications of the risk
- Platform or system supportability
- Platform or system intricacy
- The number of departments affected by the platform or system

- ■ Historical probability of failure
- ■ Risk register

Energy company D

- ■ Strategic fit
- ■ Competitive advantage
- ■ Market share increase
- ■ Time to break even
- ■ Resources
- ■ Technical complexity

Logistics and energy company A

- ■ Strategic fit
- ■ Total cost of ownership per year
- ■ Size and complexity
- ■ Risks
- ■ Dependencies on other departments
- ■ Financial (NPV)

The most popular scoring model variables are presented in Table 11.7.

Portfolio Balance

A total of seven models were employed by the five companies in the study.

Energy company A

- ■ Risk vs. reward

Table 11.7   Energy and Logistics Industry: Most Popular Scoring Variables

Variable	Frequency (%)
Strategic fit	80
Financial reward	80
Risk	60
Customer/market attractiveness	60
Technical feasibility	60
Competitive advantage	60

Table 11.8 Energy and Logistics Industry: Most Popular Balance Dimensions

Variable	Frequency (%)
Reward	86
Risk	57

Energy company B

- NPV vs. leverage core competencies
- NPV vs. commercial risk

Energy company C

- No balance model

Energy company D

- Risk vs. time to break even
- PR risk vs. time to break even
- Technical complexity vs. market attractiveness

Logistics and energy company A

- NPV vs. strategic fit

The most popular portfolio balance variables are presented in Table 11.8.

Strategic Alignment

All five companies selected the "top-down, bottom-up" approach for the strategic alignment.

Telecommunications Industry

Scoring Models

The telecommunications industry was represented by four companies with the following results:

Mobile provider A

- Financial (ROI)
- Competitive advantage

- Improve customer satisfaction
- Innovativeness
- Strategic fit
- Time to market

Mobile provider B

- Strategic fit
- Financial
- Technical feasibility
- Market attractiveness
- Resources

Mobile provider C

- Financial
- Leverage core competencies
- Innovativeness
- Simple IT architecture
- Introduce new segments
- Availability of critical resources

Mobile provider D

- Strategic fit
- Complexity (how many external resources will be required?)
- Economic impact

The most popular scoring model variables are presented in Table 11.9.

Table 11.9 Telecommunications Industry: Most Popular Scoring Variables

Variable	Frequency (%)
Financial reward	100
Strategic fit	75
Customer/market attractiveness	75
Technical feasibility	75
Resources	50

Portfolio Balance

A total of six models were employed by the four companies in this industry.

Mobile provider A

■ Risk vs. reward

Mobile provider B

■ Strategic fit vs. NPV
■ Project risk vs. NPV

Mobile provider C

■ Strategic fit vs. NPV
■ Risk vs. reward (NPV)

Mobile provider D

■ Strategic fit vs. NPV

The most popular portfolio balance variables are presented in Table 11.10.

Strategic Alignment

All four companies in the sample selected the "top-down, bottom-up" approach for the strategic alignment.

Government Sector

Scoring Models

The government sector was represented by four organizations with the following results:

Table 11.10   Telecommunications Industry: Most Popular Balance Dimensions

Variable	Frequency (%)
Reward	100
Risk	50
Strategic fit	50

Ministry of defense

- Strategic fit
- Resources required
- Financial value
- In-house knowledge/project complexity
- Risks

Federal loan and mortgage lending agency

- Project size
- Strategic fit
- Financial (internal rate of return [IRR])
- Social value
- Risks

Canadian university

- Strategic fit
- Resources required
- Technical feasibility
- Financial value
- Risks

European national bank

- Eurosystem policy alignment
- National policy alignment
- Strategic fit
- Resources
- Risks

The most popular scoring model variables are presented in Table 11.11.

Portfolio Balance

A total of 11 models were employed by the four organizations.

Ministry of defense

- Probability of success vs. impact on the customers
- Risk vs. customer value
- Risk vs. cost savings
- Risk vs. quality

Table 11.11   Government Sector: Most Popular Scoring Variables

Variable	Frequency (%)
Strategic fit	100
Risks	100
Resources	100
Financial reward	75
Technical feasibility	50

Table 11.12   Government Sector: Most Popular Balance Dimensions

Variable	Frequency (%)
Risk	82
Reward	45

Federal loan and mortgage lending agency

- Strategic fit vs. IRR
- Risk vs. IRR

Canadian university

- Risk vs. reward
- Academic vs. research vs. system support projects

European national bank

- Financial risk vs. reward
- Operational risk vs. reward
- Reputational risk vs. reward

The most popular portfolio balance variables are presented in Table 11.12.

Strategic Alignment

All four organizations selected the "top-down, bottom-up" approach for the strategic alignment.

Professional Services Industry

Scoring Models

The professional services sector consisted of four companies with the following results:

Professional services company A

- Strategic importance
- Financial forecast
- Risks
- Internal influence

Professional services company B

- Strategic fit
- Technical feasibility
- Leverage core competencies
- ROI
- Probability of winning the bid

Professional services company C

- Strategic fit
- Enable revenue growth
- Cost efficiency
- Resources required
- Time to market

The most popular scoring model variables are presented in Table 11.13.

Table 11.13   Professional Services Industry: Most Popular Scoring Variables

Variable	Frequency (%)
Strategic fit	100
Financial reward	100
Risks	33
Technical feasibility	33

Portfolio Balance

A total of six models were employed by the three companies in the study.

Professional services company A

- Financial value (ROI) vs. resources
- Financial value (ROI) vs. time to completion
- Strategic fit vs. risk

Professional services company B

- Risk vs. reward

Professional services company C

- Strategic fit vs. ROI

The most popular portfolio balance variables are presented in Table 11.14.

Table 11.14   Professional Services Industry: Most Popular Balance Dimensions

Variable	Frequency (%)
Reward	80
Risk	20

Table 11.15   Aggregate Statistics: Most Popular Scoring Variables

Variable	Frequency (%)
Financial	90
Strategic fit	87
Tech feasibility	60
Customer/market attractiveness	53
Risks	43
Resources	23
Competitive advantage	20
Synergies	17

Table 11.16   Aggregate Statistics: Most Popular Balance Dimensions

Variable	Frequency (%)
Financial reward	79
Risk	68
Others	26
Strategic fit	15
Cost	6
Resources	6

Strategic Alignment

All three organizations selected the "top-down, bottom-up" approach for the strategic alignment.

Aggregate Statistics

Scoring Models

The aggregate statistics for the most popular scoring variables across all industries are presented in Table 11.15.

Portfolio Balance

The aggregate statistics for the most popular portfolio balance variables across all industries are presented in Table 11.16.

Strategic Alignment

All 30 organizations in this study used the "top-down, bottom-up" model for the strategic alignment.

Chapter 12

Implementing Project Portfolio Management: Lessons Learned from Implementations

Introduction and Overview

Before we proceed with the tips and tricks of project portfolio management (PPM) implementation, let us revisit the entire PPM process from start to finish (see Figure 12.1).

Level 1: Project Selection Reviews

Once the scoring, balance, and strategic alignment models have been developed, the organization should proceed to the project selection review stage. The goal here is to analyze and review the project ideas by answering the following questions for each initiative:

- How do we maximize the value?
- How do we balance our portfolio?
- How do we align it with strategy?

It is also a good idea to conduct an inventory of both the financial and human resources available to support new initiatives. While financial resources are fairly

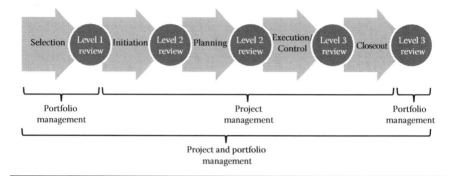

Figure 12.1 Project portfolio management lifecycle.

easy to calculate—we simply have to examine the total projects' budget amount assigned by the executives—calculation of human resources is somewhat more challenging.

Why Should You Consider Your Internal Resource Costs?

In order to analyze this problem, let us consider two scenarios. In the first one, we have to choose between projects A and B. For simplicity, let us assume that the only factor of importance in this case is the return on investment (ROI), one of the most uncomplicated financial formulas.

Project A should generate a revenue of $1,000,000. The external (direct) cost of this project is $500,000 (e.g., purchase of materials and equipment). In addition, the company should expect to "invest" about 200 person-months of its internal resources, but since we are not considering the internal employee costs, the overall impact of this factor on the total project budget is zero (see Table 12.1).

Project B is also expected to generate $1,000,000 in revenues, but the external cost is expected to be $750,000. Also, the human investment is estimated to be five person-months. But again, since the "internal employee" costs are ignored, the overall impact of this factor is also zero.

Under these conditions, which project is preferable? A simple calculation will tell us that project A is far more attractive than project B:

$$\text{ROI of A} = (\$1,000,000 - \$500,000)/\$500,000 = 100\%$$

$$\text{ROI of B} = (\$1,000,000 - \$750,000)/\$750,000 = 33\%$$

But what happens if we decide to incorporate the employee cost into the equation? My personal experience, based on numerous interactions with chief financial officers, suggests that in the developed countries, the average blended monthly

Table 12.1   Project Feasibility Calculation without Human Cost

	Project A	Project B
Projected revenue ($)	1,000,000	1,000,000
Direct cost (e.g., equipment and outsourcing) ($)	500,000	750,000
Human cost (man-months)	200	5
Human cost ($)	0	0
Total cost ($)	500,000	750,000
ROI (%)	**100**	**33**

Table 12.2   Project Feasibility Calculation with Human Cost

	Project A	Project B
Projected revenue ($)	1,000,000	1,000,000
Direct cost (e.g., equipment and outsourcing) ($)	500,000	750,000
Human cost (man-months)	200	5
Human cost ($)	2,000,000	50,000
Total cost ($)	2,500,000	800,000
ROI (%)	**−60**	**25**

employee cost is around $10,000. This (surprisingly high to some people) number includes salary, benefits, and employment taxes, as well as hiring, equipment, and space costs.

Let us recalculate the financial feasibility numbers for our candidate projects once more, but this time we will consider internal human resources costs (see Table 12.2):

$$\text{ROI of A} = (\$1,000,000 - \$2,500,000)/\$2,500,000 = -60\%$$

$$\text{ROI of B} = (\$1,000,000 - \$800,000)/\$800,000 = 25\%$$

So, what conclusions can we make from these two mini case studies?

- Internal employee efforts on your projects should never be viewed as "free resources."

■ Organizational leaders should make an effort to calculate at least an approximate blended monthly (daily, weekly) of employee costs.
■ The employee costs should always be included in the project feasibility calculations.

How to Establish the Size of Your Project Resource Pool

Here is an example of a "back of the envelope" calculation of total project resources bucket at a company. Imagine that there are 250 employees working at the head office. It has been estimated via survey or questionnaires that approximately 30% of their time is spent on the project work and 70% on business as usual, that is, normal daily nonproject tasks. Based on that information, we can assess the size of the total project resource bucket at the company:

Total number of people at the head office = 250 people
Total number of working months in a year = 10 minus 2 months for vacation, holidays, and sick days
Percentage of time spent on projects = 30% (estimated based on surveys)

Thus,

$$\text{Total project resource pool} = 250 \text{ people} \times 10 \text{ months} \times 0.30$$

$$= 750 \text{ person-months}$$

Therefore, the total project human resources available for the entire portfolio are 750 person-months. Using this figure and knowing that there are 12 months in a year, we can calculate the approximate resource pipeline throughput at the company as follows:

$$\text{Project pipeline capacity} = \text{Total project resources/Number of months in a year}$$

$$= 750 \text{ person-months/12 months}$$

$$= 62.5 \text{ person-months/month}$$

In other words, the total project resource requirements at the organization should not exceed 62.5 person-months in any given month.

Level 2: Phase-Level Reviews

Level 2 reviews happen at the end of the initiation and planning phases of the project. The key premise for these reviews is this:

> Now that we have completed the initiation phase and discovered a new information regarding project scope, schedule, budget, etc., with

a progressing degree of accuracy—recorded in the project charter and the project plan—is there any new information that impacts our previous decision to accept this project?

As a result, here are the questions that need to be asked at Level 2 reviews:

■ Is the original business case for the project still supported?
■ Are there any drastic changes to the following?
  – Project budget
  – Project duration
  – Project human resource requirements
  – Revenue projections
  – Other factors considered at selection such as internal and external risks and organization strategies and goals

If there were significant negative changes in any of the aforementioned categories, the project would have to be postponed, killed all together, or drastically adjusted to address the issues.

Level 3: Periodic Project Status Reviews

The periodic project status reviews usually start at the execution phase of the project and mainly focus on the answers to the following questions:

■ Is the project still on time?
■ Is the project still on budget?
■ Are there any major scope changes?
■ What are the key milestones that we passed?
■ What are the key milestones ahead?
■ Are there any unexpected technical and design issues?
■ Are there any other unexpected risks?

The key assumption that is made here is there is a low probability—assuming the initiation and planning stages have been handled properly—of any drastic changes to the original business case for the project, and we are focusing more on the tactical issues at hand.

Importance of Mission and Strategy

The Mission

Any portfolio exercise should start with a thorough discussion and analysis of the organization's mission and strategy. Why would this be one of the most important first steps? As was shown in the previous chapters, the vast majority of organizations add strategic fit variable to their scoring models when prioritizing their project proposals. In addition, a certain percentage of them prefer to monitor their portfolio balance using the "strategic fit" as one of the variables in the bubble chart diagrams. Finally, strategic alignment plays a major role in determining the size of organizational project buckets during the portfolio alignment exercise.

So, a clearly articulated and simple list of your organization's strategies would become a first major input into the portfolio management exercise. However, creation of the strategies is impossible without knowing the organizational mission.

One of the most frequently encountered problems in the author's professional practice is that many organizations choose to formulate their missions and strategies using generic, bland, and overused words including "cutting edge," "industry leader," "state of the art," "innovation," and "creativity." Unfortunately, while these expressions sound nice in sales and marketing pitches, they are too ambiguous to serve as a foundation for future decisions, activities, and projects. To illustrate this point, we can examine the following exchange from a popular 1997 movie "Men in Black":

> **James Edwards:** [who has just arrived at the MIB headquarters] Maybe you already answered this, but, why exactly are we here?
> **Zed:** [noticing a recruit raising his hand] Son?
> **Second Lieutenant Jake Jenson:** Second Lieutenant, Jake Jenson. West Point. Graduate with honors. We're here because you are looking for the best of the best of the best, sir!
> **Zed:** [throws Edwards a contemptible glance as Edwards laughs] What's so funny, Edwards?
> **James Edwards:** Boy, Captain America over here! "Best of the best of the best, sir!" "With honors." Yeah, he's just really excited, and he has no clue why we're here.

Therefore, to avoid looking like Second Lieutenant Jake Jenson, the executives must word their mission statements—as well as goals and strategies—in a way that is specific, measurable, and relevant.

In order to clarify this statement, let us examine a real-life example of a dialogue with the R&D executives of ball bearings manufacturer.

> Me: So, what is your mission?
> Executives: Well, we want to be a global leader in innovation and creativity by delivering the best mix of products and services to our clients worldwide
> Me: Sorry, but this mission statement is applicable to pretty much any business out there… Can you be a bit more specific?
> Executives: You see, we have been in the business of developing and manufacturing of ball bearings for a while. We now think that we

should expand into other related products. We want to get there fast, and we want to offer competitive stuff...

Me: Then how about this: "XYZ Company will achieve a X% market share in sealants, lubricants, and electronic components markets while maintaining its Y% in the existing ball bearings market?"

Executives: That sounds much better...

Company's Strength, Weaknesses, Opportunities, and Threats Analysis

Conducting a SWOT (strengths, weaknesses, opportunities, and threats) analysis can ease the transition from the organization's mission statement to the list of strategies.

Here is a simple list of questions one needs to find the answers in this exercise:

- What are your (internal) strengths?
- What are your (internal) weaknesses?
- What are the (external) opportunities you have?
- What are the (external) threats you are facing?

Let us examine this model using an example. The first one is the bearings manufacturer mentioned earlier in this chapter.

Q: What are our internal strengths?
A: We have been very successful producing all types of bearings for several decades. Our brand name is recognized and respected for the value and quality of our products. We are enjoying a dominant position in the market.

Q: What are our internal weaknesses?
A: Our customers have been inquiring a lot about other bearings-related products, including sealants, lubricants, and electronic components. Unfortunately we do not have these products in our portfolio.

Q: What external opportunities do we have?
A: We can considerably grow our revenues by adding sealants, lubricants, and electronic components to our product portfolio.

Q: What are the external threats we are facing?
A: Our competitors have been offering complement products for a while now, slowly eroding our market share.

Goals and Strategies

When determining the organization's goals, the executive management must examine the strategies logically resulting from both the mission statement and the SWOT analysis. Some of the questions they may ask include the following:

- What are our five-year goals?
- What is our goal for this year?
- What are our three-month goals?

Let us return to our example discussed earlier. As a result of an analysis of both the mission statement and the SWOT questions and answers, the bearings manufacturer executives came up with the following goals:

> Q: What are our five-year goals?
> A: We want to capture a large market share of products complementary to bearings, namely, sealants, lubricants, and electronic components. We also want to continue remaining the market leader in bearings.
> Q: What are our goals for this year?
> A: We want to establish proper project portfolio management (PPM) processes to select the best "complement" projects. We need to redirect our R&D resources into the new products.
> Q: What are our three-month goals?
> A: We need to start looking at the new proposals aimed at developing the new sealants, lubricants, and electronic components.

Why Do You Need Direct Executive Involvement?

Another seemingly simple but frequently encountered illusion is that the PPM process can somehow go ahead without the direct involvement of the executive management. Very frequently when teaching my public PPM masterclass, I am engaged in the following conversation with my "students":

> S: Hi, my name is Pascal, and I am a newly appointed Director of Portfolio Management at company X. My goals today are to learn as much as possible about practical PPM and implement it once I get back to our headquarters.
> Me: Will your executive management participate in the process?
> S: No, unfortunately not. They are very busy people, you know… but they assured me I would have their full support in this undertaking!
> Me: So, who is going to create the scoring matrix, balance, and strategic alignment models?

S: Hopefully I will do that once I am done with the course.

Me: Understood. Let me paint the following picture for you, and you can tell me at the end whether this sounds as a plausible scenario: you create the scoring matrix, balance, and strategic alignment models. At one point of time you are approached by your company's CEO or senior VP who asks you to add this "very important project" to the organizational portfolio and initiate it as soon as possible. You however upon analyzing the proposal come back to your senior executive and tell her that the project will not be going ahead because it scored only five out of the possible 50 points in the prioritization model. What do you think her response will be?

S: She will ask me who created the model!

Me: And once you reply that the model was your creation, what is the likely response?

S: She will probably say something to the effect of that the creation of such models wasn't really in my domain of responsibilities.

Me: Aha! And we haven't even touched the questions regarding company's mission and strategy that ties directly into the formation of the portfolio! Now you see why executive involvement is absolutely essential?

In other words, what credibility does the ranking criteria created by the Director of Portfolio Management have in the eyes of the CEO or of any other executive of the company? We can expand this argument even further and ask what value a model created by the CEO alone will hold in the eyes of all other organizational executives if they were not allowed to participate in its creation.

Another important factor that should be considered when organizing project portfolio reviews is the fact that having managers from all domains within the organization eliminates the possibility of mistakes or skew toward one particularly powerful executive or department. The problem is that the executives—just like the rest of us—tend to suffer from something known as "optimism bias." Applied to PPM, this phenomenon can be defined as follows:

The managers tend to overestimate the value of the projects they propose and/or underestimate the complexity (e.g., budgets, resources, timelines, risks) of the said ventures.

There are three potential explanations for this fact:

1. The "cognitive dissonance" theory
2. The "mass delusion" theory
3. The "Machiavelli factor" theory

"Cognitive Dissonance" Theory

The cognitive dissonance theory was developed by Leon Festinger in 1956 and states the following:

- People have difficulties holding inconsistent beliefs.
- Hence, they gladly pick up the information that supports their beliefs and readily reject that which does not.
- It occurs unintentionally, and we rarely have any control of these processes in our brain.
- Therefore, if the executive gave her approval to the project, she has implicitly expressed her belief in the value and the "goodness" of the venture.
- However, if the project proceeds in an unplanned way for either strategic (PPM) or tactical (project management) reasons, she will subconsciously have a hard time accepting the truth.
- As a result, she will continue believing in the positive information about the venture, but she will reject the bad news received from the project team.

"Mass Delusion" Theory

The key experts in what is called "behavioral economics," Dan Lovallo and the 2002 Nobel Prize winner Daniel Kahneman propose the following explanation in their article "Delusions of Success: How Optimism Undermines Executives' Decisions" (Lovallo and Kahneman, 2003):

- What happens in today's business world has little to do with calculated business risks.
- Modern business decision making is seriously flawed because of the delusional optimism (optimism bias) that forces people to overestimate the benefits and underestimate the costs of future projects.
- Business executives tend to exaggerate the degree of control that they have over events, discounting the role of luck.
  - For example, a multi-industry study of start-ups found that more than 80% failed to achieve their market share target.
- Business leaders routinely exaggerate their personal abilities, particularly for ambiguous, hard-to-measure traits such as managerial skill.
- Another factor called anchoring also prevents us from producing accurate estimates. Simply put, when the executives propose the new project, they tend to accentuate the positives in order to make the case for their proposal. As a result, all future estimates are skewed toward overoptimism. This phenomenon is the result of anchoring, one of the strongest and most prevalent of cognitive biases.

    – For example, one Rand Corporation study of 44 chemical-processing plants owned by major companies, such as 3M, DuPont, and Texaco, found that on average the factories' actual construction costs were more than double the initial estimates. Furthermore, even a year after start-up, about half the plants produced at less than 75% of their design capacity, with a quarter producing at less than 50%.

The conclusion from Lovallo and Kahneman as a result of these findings is that companies must employ external experts to generate accurate and unbiased estimates for their projects.

"Machiavelli Factor" Theory

Professor Bent Flyvbjerg (2003), who has dedicated his career to the study of ambitious megaprojects, disagrees both with the "cognitive dissonance" and the "mass delusion" theory explanations. In his article titled "Delusions of Success: Comment on Dan Lovallo and Daniel Kahneman," he mentions the following counterarguments:

- In the course of his studies, he and his colleagues frequently encountered the deliberate "cooking of the books" to make the project proposal look more attractive. He calls it the "Machiavelli factor."
- His analysis of capital transportation projects revealed that the executives were rewarded with large incentives for positive forecasts and faced only minor penalties when their predictions proved to be wrong.
- Moreover, Professor Flyvbjerg contends that during the course of 70 years covered by his study, the relative size of the estimation errors remained suspiciously constant.
- For example, urban rail investments were on average 45% over budget, while the actual ridership was 50% lower than predicted.
- Finally, he argues that since humans are—as a rule—quite capable of learning from their mistakes, it is unlikely that they would continue to make the same mistakes decade after decade.

Scientific studies conducted by Bent Flyvbjerg show that political pressure is the top influencer, hence the C-level would be unwilling to accept external fair estimates.

Project Portfolio Management Charter

It is advisable for organizations to create and institutionalize project and portfolio management charters to cover the following topics:

- Clarify the roles, responsibilities, and expectations of all the project stakeholders including the executive committee, departmental managers, project management office (PMO)/PPM office managers, project managers, and internal customers.
- Describe all project and portfolio management–related organizational processes and procedures including creation, scoring and approval of the business cases, project initiation, planning, execution/control, and closeout stages.
- Establish the communication procedures and channels (e.g., portfolio committee with the department heads, project managers with the portfolio committee, portfolio committee with the organization's employees).
- Include the current portfolio scoring model, balance, and the strategic fit with clear and measurable criteria for each of the scoring ranges (e.g., "1 point," "5 points," and "10 points").
- Establish portfolio cycles and checkpoints at the end of each of the initial project stages and regular "gates" during the project execution phase.
- Defines project and portfolio performance metrics.

Table 12.3 contains a sample table of contents of this document and the topics it covers.

Portfolio Scoring Model and Project Ranking

Halo Effect

When developing project portfolio scoring models, one should be aware of the halo effect that can have a serious impact on the final ranking results. This effect happens when the scoring model developers select variables that are closely correlated with one another, thus skewing the overall performance of the model.

Here is an obvious example of the halo effect in action. The executive committee, heavily dominated by the people with financial and accounting backgrounds, proposed the following model:

- Strategic fit
- ROI
- Market attractiveness
- Net present value (NPV)
- Payback
- Technical feasibility

Do you see anything peculiar about this model? Three out of the six variables proposed were of a financial nature (i.e., ROI, NPV, and payback). It is easy to see that they would exhibit a strong correlation with each other. In other words,

Table 12.3   Sample Table of Contents

TABLE OF CONTENTS
REVISION HISTORY TABLE
PROJECT AND PORTFOLIO GOVERNANCE CHARTER PURPOSE
DEFINITIONS AND VALUE OF PROJECT MANAGEMENT AND PORTFOLIO MANAGEMENT
Project Management
Project Portfolio Management
OUR COMPANY'S PROJECT AND PORTFOLIO MANAGEMENT ROLES AND RESPONSIBILITIES
OUR COMPANY'S PROJECT AND PORTFOLIO MANAGEMENT PROCEDURES OVERVIEW
Our Company's Project Management
Our Company's Project Portfolio Management
Pre-Initiation (Business Case)
DETAILED PROJECT MANAGEMENT PROCEDURE
Initiation
Planning
Execution and Control
Close-Out
DETAILED PROJECT PORTFOLIO MANAGEMENT PROCEDURE
Portfolio Value
The Joker Project Concept
Total Project Resource Pool Estimates
Portfolio Balance
Portfolio Strategic Alignment

(Continued)

Table 12.3 (*Continued*) Sample Table of Contents

PROJECT MANAGEMENT AND PROJECT PORTFOLIO MANAGEMENT KEY PERFORMANCE INDICATORS
TEMPLATES LIBRARY
SUGGESTIONS FOR IMPROVEMENT
REFERENCES AND ATTACHMENTS
References
Attachments
GLOSSARY

project proposals with high NPV will tend to have higher IRR and a shorter payback period.

As a result, when this model was applied to the first several project proposals, it became painfully obvious that the company portfolio was dominated by smaller and simpler (because of the technical feasibility factor) projects that promised a high ROI (or higher NPV or a shorter payback time). However, all large strategic initiatives as well as "stay-in-business" maintenance projects dropped to the bottom of the portfolio and had to be saved by invoking the "joker" powers.

Consequently, managers were advised to recalibrate the model by uniting IRR, NPV, and payback into one "financial performance" variable and transform the scoring model to

- Strategic fit
- Financial performance (a combination of ROI, NPV, and payback)
- Market attractiveness
- Technical feasibility

Another more subtle example of the halo effect involves an IT department of a North American university. The committee dominated by the IT professionals developed the following scoring model:

- Strategic fit
- Size and complexity
- Risks
- Dependencies on other departments
- Financial (NPV)
- Market demand

Although at the first glance the model looks fine, a more thorough examination reveals that the variables "size and complexity," "risk," and "dependencies on other departments" are bound to be highly correlated:

■ The larger the project, the more likely it is that it would involve more departments
■ The more complex the project, the more likely that it would be riskier

As a result of this decision, the portfolio was dominated by smaller and simple system enhancement ventures, while all major implementations and upgrades fell to the bottom of the project stack. Thus, the company management was forced to replace these three variables with one composite factor called "project complexity and risk," which was a function of project size and complexity, riskiness, and a number of departments involved. The final scoring model looked as follows:

■ Strategic fit
■ Project complexity and risk
■ Financial (NPV)
■ Market demand

Project Proposal, aka the Business Case

The business case document should answer the following question:
Should we do this project and why?
The answers available to the managers therefore are as follows:

■ Accept this project.
■ Kill this project.
■ Postpone this project and rework its scope (and/or timeline, and/or budget) in order to make it acceptable.

The business case should align with the company's scoring model and include the following sections (see Tables 12.4 and 12.5 for a template and a sample document):

■ *Name of the proposed project*—A fairly straightforward part where the proposer or sponsor should attempt to come up with a short but descriptive title for the initiative being proposed.
■ *Sponsor's contact information*—The proposer's first name, last name, position within the organization, phone number, and e-mail address.
■ *Description of the project scope*—Two or three paragraphs describing the overall project scope and the deliverables. In addition, if possible, the proposer should include a feature-by-feature list of all the key ingredients in the project's scope.

Table 12.4 Sample Business Case Template

Project Name					
			Min	Max	
Forecasted Project Budget					
Forecasted Project Resource Requirements (Man-Months)					
Forecasted Project Duration					
Selection Criteria		*Points Awarded*			
Must do or "joker project" A strategic "must do" initiative that has to be implemented irrelevant of all other selection criteria and must be approved by the executive-level management		61 points			<Explain why this project falls into the "joker" category>
	1	5	10		
Strategic fit Does the proposed project fit one or more of these criteria?	Low Fits one or two of the criteria	Medium Fits three or four of the criteria	High Fits five or more of the criteria	<Select the appropriate number of points for this category from the drop-down menu>	<Explain how many strategies does this project fit>

(Continued)

Table 12.4 (*Continued*) Sample Business Case Template

Project Name				
Forecasted Project Budget			*Min*	*Max*
Forecasted Project Resource Requirements (Man-Months)				
Forecasted Project Duration				
Selection Criteria	*Points Awarded*			
1. Increase the number of SME clients.				
2. Increase the number of big business clients.				
3. Decrease the size of credit per existing customer (by increasing the customer base).				
4. Increase portfolio diversification.				
5. Improve bad debt management.				
6. Improve credit risk management.				
7. Increase the number of new products (including fees and commissioning).				

(Continued)

Table 12.4 (Continued) Sample Business Case Template

Project Name				
			Min	Max
Forecasted Project Budget				
Forecasted Project Resource Requirements (Man-Months)				
Forecasted Project Duration				
Selection Criteria	*Points Awarded*			
Note: This is a kill category. If the proposed project scores 0 (zero) on a strategic fit criterion, it is removed from further consideration.				
ROI What is the expected ROI for the proposed project?	Low <10%	Medium 10%–12%	High 12+%	<Select the appropriate number of points for this category from the drop-down menu.>	<Describe how the ROI number has been arrived at. Show expected future cash flows versus expected expenses.>

(Continued)

Table 12.4 (Continued)   Sample Business Case Template

Project Name		
	Min	*Max*
Forecasted Project Budget		
Forecasted Project Resource Requirements (Man-Months)		
Forecasted Project Duration		

Selection Criteria	Points Awarded				
	Low	Medium	High		
Market attractiveness/competitive advantage How high is the market demand for this project? Are there many competitors offering same product or service?	Market demand is low. Many competitors are offering similar products or services	Market demand is medium. Some competitors are offering similar products or services	Market demand is high. Few or none of the competitors are offering similar products or services	<Select the appropriate number of points for this category from the drop-down menu.>	<Provide relevant information regarding the demand for the product, e.g., how many direct customer requests have been logged. Outline how many competing banks have been offering the same product or service.>

Table 12.4 (*Continued*) Sample Business Case Template

Project Name			
		Min	Max
Forecasted Project Budget			
Forecasted Project Resource Requirements (*Man-Months*)			
Forecasted Project Duration			
Selection Criteria	Points Awarded		

Selection Criteria	Points Awarded			
	Low	Medium	High	
In-house expertise	The bank does not have internal experts	The bank has some internal experts	The bank has a lot of internal experts	<Select the appropriate number of points for this category from the drop-down menu>
Does the bank has internal expertise to deliver this project? Will it require a lot of external resources?	The project will require a lot of external expertise	The project will require some external expertise	The project will require little or no external expertise	<Explain whether a lot of external expertise will be required in terms of vendors, external consultants, etc.>

(*Continued*)

Table 12.4 (*Continued*) Sample Business Case Template

Project Name				
			Min	Max
Forecasted Project Budget				
Forecasted Project Resource Requirements (Man-Months)				
Forecasted Project Duration				
Selection Criteria	Points Awarded			
Risk and complexity Will this project encounter a lot of risks (e.g., technical, organizational, operational, legal, and compliance) during its delivery? How complex is this project from the technical point of view? Note: This is a kill category. If the proposed project carries substantial compliance or legal risks, it is removed from further consideration.	High A lot of risks are expected to surface during the project Very complex project	Medium Some risks are expected to surface during the project Somewhat complex project	Low Few or no risks are expected to surface during the project Simple project	<Select the appropriate number of points for this category from the drop-down menu.>	<Provide relevant information regarding the potential risks of the project. Also, provide information regarding the perceived complexity of the project.>

(Continued)

Table 12.4 (Continued) Sample Business Case Template

Project Name				
Forecasted Project Budget			Min	Max
Forecasted Project Resource Requirements (Man-Months)				
Forecasted Project Duration				
Selection Criteria	Points Awarded			
Improves operational efficiency? Will this project improve the operational efficiency at PASHA Bank? How significant is the expected impact?	Low Little or no expected positive impact on the operational efficiency	Medium Some expected positive impact on the operational efficiency	High Significant expected positive impact on the operational efficiency	*<Select the appropriate number of points for this category from the drop-down menu.>*	*<Describe the expected impact of the project on the operational efficiency. Will it decrease the time spent on specific tasks, improve the ability of PASHA Bank employees to perform specific activities, decrease the time to market, etc.?>*
			Total score	

Table 12.5   Sample Business Case: Opening of a New Branch in City A

Project Name: Opening of a New Branch in City A		Min	Max	
Forecasted Project Budget		*$500,000*	*$1,000,000*	N/A
Forecasted Project Resource Requirements (Man-Months)		*100*	*400*	
Forecasted Project Duration		*9 Months*	*18 Months*	
Selection Criteria	*Points Awarded*			
Must do or "joker project" A strategic "must do" initiative that has to be implemented irrelevant of all other selection criteria and must be approved by the executive-level management	61 points			
	1	**5**	**10**	
Strategic fit Does the proposed project fit one or more of these criteria:	Low Fits one or two of the criteria	Medium Fits three or four of the criteria	High Fits five or more of the criteria	The project covers strategies (1), (2), (3), (4), and (6)
			10	

(Continued)

Table 12.5 (*Continued*)   Sample Business Case: Opening of a New Branch in City A

Project Name: Opening of a New Branch in City A		Min	Max	
Forecasted Project Budget		$500,000	$1,000,000	
Forecasted Project Resource Requirements (Man-Months)		100	400	
Forecasted Project Duration		9 Months	18 Months	
Selection Criteria	Points Awarded			
1. Increase the number of SME clients.				
2. Increase the number of big business clients.				
3. Decrease the size of credit per existing customer (by increasing the customer base).				
4. Increase portfolio diversification.				
5. Improve bad debt management.				
6. Improve credit risk management.				
7. Increase the number of new products (including fees and commissioning).				

(Continued)

Table 12.5 (Continued) Sample Business Case: Opening of a New Branch in City A

Project Name: Opening of a New Branch in City A		Min	Max	
Forecasted Project Budget		$500,000	$1,000,000	
Forecasted Project Resource Requirements (Man-Months)		100	400	
Forecasted Project Duration		9 Months	18 Months	
Selection Criteria		Points Awarded		
Note: This is a kill category. If the proposed project scores 0 (zero) on a strategic fit criteria, it is removed from further consideration.				
ROI What is the expected ROI for the proposed project?	Low <10%	Medium 10%–12%	High 12+%	10	Expected ROI is around 20%–25% (see attached document for financial calculations)

(Continued)

Table 12.5 (Continued)   Sample Business Case: Opening of a New Branch in City A

Project Name: Opening of a New Branch in City A			
		Min	Max
Forecasted Project Budget		$500,000	$1,000,000
Forecasted Project Resource Requirements (Man-Months)		100	400
Forecasted Project Duration		9 Months	18 Months
Selection Criteria	Points Awarded		
Market attractiveness/competitive advantage	Low	Medium	High
How high is the market demand for this project? Are there many competitors offering same product or service?	Market demand is low	Market demand is medium	Market demand is high
	Many competitors are offering similar products or services	Some competitors are offering similar products or services	Few or none of the competitors are offering similar products or services
			5
			Currently there are only two other banks that have representative branches in city A: banks X and Y
In-house expertise	Low	Medium	High
Does the bank has internal expertise to deliver this project? Will it require a lot of external resources?	The bank does not have internal experts	The bank has some internal experts	The bank has a lot of internal experts
			5
			Construction of the building—outsourced

(Continued)

Table 12.5 (Continued) Sample Business Case: Opening of a New Branch in City A

Project Name: Opening of a New Branch in City A			Min	Max	
Forecasted Project Budget			$500,000	$1,000,000	
Forecasted Project Resource Requirements (Man-Months)			100	400	
Forecasted Project Duration			9 Months	18 Months	
Selection Criteria	Points Awarded				
Risk and complexity	The project will require a lot of external expertise	The project will require some external expertise	The project will require little or no external expertise		All other scope items—handled in-house
Will this project encounter a lot of risks (e.g., technical, organizational, operational, legal, and compliance) during its delivery? How complex is this project from the technical point of view?	High A lot of risks are expected to surface during the project	Medium Some risks are expected to surface during the project	Low Few or no risks are expected to surface during the project	10	Low-risk project due to previous experienced on similar branch-opening projects

(Continued)

Table 12.5 (Continued) Sample Business Case: Opening of a New Branch in City A

Project Name: Opening of a New Branch in City A

			Min	Max	
Forecasted Project Budget			$500,000	$1,000,000	
Forecasted Project Resource Requirements (Man-Months)			100	400	
Forecasted Project Duration			9 Months	18 Months	
Selection Criteria		*Points Awarded*			
Note: This is a kill category. If the proposed project carries substantial compliance or legal risks, it is removed from further consideration.	Very complex project	Somewhat complex project	Simple project		
Is operational efficiency improved? Will this project improve the operational efficiency at PASHA Bank? How significant is the expected impact?	Low Little or no expected positive impact on the operational efficiency	Medium Some expected positive impact on the operational efficiency	High Significant expected positive impact on the operational efficiency	1	Some positive effect on the operational efficiency. Account managers will not have to travel to city A to meet with clients
			Total score	41/50	

- *Discussion of the portfolio fit*—The prefilled scoring matrix, the type of project, and, unless the topic has been covered in the scoring matrix, the strategic fit of the project.
- *Estimates*—Verify the project scoring model to see if it includes financial and/ or resource variables. In addition, these data are significant when the executives are going through the strategic bucket filling exercise during the portfolio assembly (see the section "How It All Works in Real Life" in Chapter 2).
- Since the generation of the precise (e.g., ±10%) estimate is not only impossible, but also misleading at this stage, the organization should decide on the acceptable ranges. For example,
  - Familiar projects: +75%; –25%
  - New ventures: +300%; –75%

  Alternatively, if the organization decided to use the PERT methodology for the strategic resource allocation exercises, the proposal author should include the pessimistic, most likely, and optimistic estimates for the required categories.

  Finally, if the project requires any special resources that are in a short supply or need to be procured externally, they should be part of the proposal document.
- *Risks*—Risks are usually defined as negative things that can happen on your project, but you are not entirely sure they will happen. For example, the most common risks that happen at this stage include underestimation of the scope size and complexity, need for external expertise, missing requirements, and optimism bias in the estimate generation, just to name a few.

Try to Generate as Many Project Proposals as Necessary

A great American scientist and the 1954 Nobel Prize winner Linus Pauling in Chemistry once said, "If you want to have good ideas you must have many ideas. Most of them will be wrong, and what you have to learn is which ones to throw away."

One of the key principles of PPM is to generate as many project proposals as possible to improve the portfolio selection. The idea here is the more proposals one has in front of him or her, the easier it would be to build a higher-quality portfolio of projects. Let us try to illustrate that point with a couple of examples (see Figures 12.2 and 12.3).

Figure 12.2 shows the bubble chart diagram of several project proposals using the risk–reward diagram. Let us assume that we need to select six projects out of all the proposals in front of us. In this particular example, once we eliminate the always unwanted "white elephants," that is, the low-reward and high-risk projects, we have exhausted our choices. In other words, we must add all of the remaining ventures to our portfolio including two "pearls," two "bread and butter," and two "oysters."

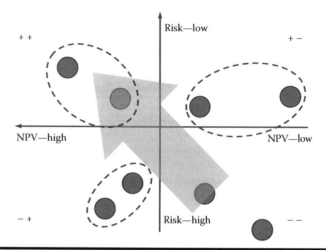

Figure 12.2   Portfolio balance—few project proposals.

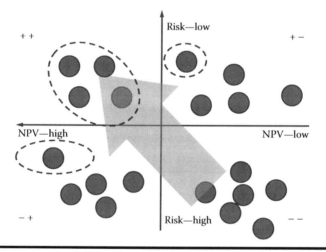

Figure 12.3   Portfolio balance—many project proposals.

Let us compare this situation with the portfolio shown in Figure 12.3. Instead of having only eight candidates, we now have 20: four "pearls," five "bread and butter," five "oysters," and six "white elephants."

After eliminating all the low-reward, high-risk projects and an addition of four low-risk, high-reward ones, we are now free to mix and match the "oysters" and the "bread and butters" to our liking. For example, we may decide to select two highest-return and lowest-risk projects from the "bread and butter" and "oyster" quadrants, respectively.

Comparing these two charts clearly demonstrates that having more project ideas to start with we are able to select a more attractive portfolio overall.

How to Collect the Largest Number of Proposals Possible

One of the first steps in collecting the maximum possible number of proposals is to make the submission process simple. One of the approaches used by some of my clients was to publish the finalized version of the scoring model on the company's Intranet that was accessible to all company employees. Another organization printed the scoring model on large posters and placing them around their company's office, including the offices of the key decision makers (see Table 12.6).

Some other valuable strategies for maximizing the number of project proposals include

- Allowing your employees to participate in the preliminary ranking processes
- Rewarding your workers financially for successful ideas
- Running contests for the best project ideas
- Inviting your customers, vendors, and suppliers to participate in the process
- Making the proposal submission process as easy and transparent as possible
- Demonstrating the positive impact on sales
- Publishing the final approved project list with the scores that each proposal has generated (see Table 12.7)

Portfolio Monitoring

According to the research by Marakon Associates in 2005 (Michael, 2005), only 15% of organizations track project performance against benchmarks. This basically implies that once the proposal receives a "go ahead" from the executive committee, it becomes a "proverbial runaway train"; no one bothers to stop it at a specific station in order to ensure that whatever assumptions and predictions made at the very inception of the project still hold true.

Therefore, it is necessary to highlight and describe in detail the importance of the project management and the PMO in the success of the PPM.

Sound Project Management Capabilities Are Essential

One of the most crucial prerequisites of the portfolio management implementation success is having a structured, centralized project management methodology in place. In their article titled "Why CEOs Fail," R. Charan and G. Colvin (1999) state the following:

- Seventy percent of the CEOs fail not because of poor strategy but because of poor execution.
- A study of 200 companies in the United Kingdom found that 80% of company directors felt they had the right strategy, but only 14% believed that these strategies were implemented properly.

Table 12.6 Sample Scoring Model Poster

Selection Criteria	Points Awarded			
		91 Points		
Joker	*1 Point*	*5 Points*	*15 Points*	*Kill?*
Strategic fit Grow canning business by developing new canning products and solutions. Cut operational costs. Focus on geographical areas A, B, C, and D.	Low Fits only one of our strategies	Medium Fits two of our strategies	High Fits all three of our strategies	Yes If does not fit any of the strategies and not a regulatory or "joker" project
Time to market	High T > 5 years	Medium 2 < T < 5 years	Low T < 2 years	No
Market attractiveness	Poor Low expected volume sales and/or does not include "must win" customers	Medium Medium expected volume sales and/or may include "must win" customers	Excellent High expected volume sales and/or most likely will include "must win" customers	Yes If very low, unless a regulatory or "joker" project

(Continued)

Table 12.6 (*Continued*) Sample Scoring Model Poster

Selection Criteria		Points Awarded			
			91 Points		
Joker	*1 Point*	*5 Points*		*15 Points*	*Kill?*
Technical feasibility	Low Complex project involving a lot of external expertise	Medium Somewhat complex project with certain degree of outsourcing involved		High Relatively simple project, no or little outsourcing required	No
Competitive advantage	Low More than four competitors offering similar products Low probability of getting a patent	Medium Between two and three competitors offering similar products Medium probability of getting a patent		High Between zero and one competitors offering similar products High probability of getting a patent	No
Fit to existing supply chain	Low Major changes to the existing supply chain will be required	Medium Some changes to the existing supply chain will be required		High No or very few changes to the existing supply chain will be required	No

Table 12.7 Sample Project List

Project Name	Score	Comments
Credit risk rating	61/60	Regulatory joker project; mandated by the central bank
Core banking system replacement	61/60	Joker project; mandated by the executive committee
New data center	61/60	Joker project; mandated by the executive committee
Open branch in city A	55/60	
Open branch in city B	51/60	
Risk management system upgrade	46/60	
Product A for SMEs	45/60	
Product B for SMEs	41/60	

To understand the real role of project management in the establishment of the PPM, let us revisit yet again the entire process in depth (see Figure 12.4).

The project starts as an idea by a company's employee—more often than not someone senior—and not being a technical expert, he or she has to discuss this conception with a project manager to obtain high-level scope, budget, and duration estimates. At this point of time, the conversation between the manager (M) and the project manager (PM) may look something like this:

> M: Hey John, we have been thinking about opening a new store in Paris. Don't have much information about it, but could you come up with a ballpark estimate for this initiative?
> PM (after doing some homework on the subject): Well, according to our historical data this project would cost anywhere between US$ 1 and 2 million and include the following high-level features…

The manager would go away, write the business case for this project proving the feasibility of the new store in France, and present it to the executive committee. The committee would either approve or reject the proposal. If the initiative is approved, the manager would need to work with the project manager and try to get more accurate information regarding the project scope, timeline, and budget for the Level 2 reviews at the end of the initiation and planning stages.

When the project enters the execution/control stages, the project champion continues to receive critical information from the project manager regarding the project scope, timeline, and budget. To make a long story short, portfolio management

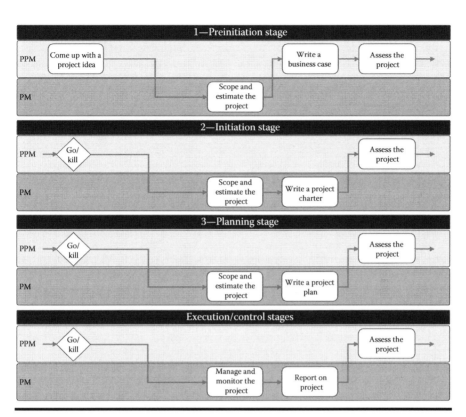

Figure 12.4 Project portfolio management lifecycle—detailed view.

is a process of constant interaction between PPM and project management; the information is constantly flowing in both directions, and the key decisions are being made based on those data.

Role of the PMO

The role of the PPM process can be generally divided into two domains: project prioritization and selection as well as project management and monitoring (see Figure 12.5).

The first role of the PMO is to act as a filtration mechanism for all the incoming project proposals. It is important to point out that the PMO should not have a mandate to overturn or reject project requests. Its role is to accept the business cases, review them, and whenever possible to point out inconsistencies or imperfections in these documents to the project champions.

First, the prerogative of project acceptance or rejection belongs only to the portfolio executive committee consisting of the organization's executives. They are the

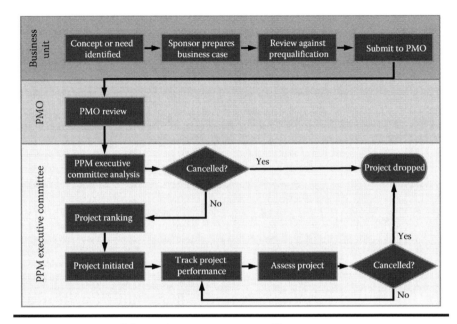

Figure 12.5   Role of the project management office.

only people who have the right to approve, reject, postpone, or kill any project in the proposal pipeline.

Second, the role of the PMO is to work with the project champions in order to guide them in writing and "selling" their project proposals. Here is an example of a conversation that may take place between the PMO manager and the project champion:

> M: John, I looked over your business case for the CRM system replacement and overall it looks very good. My only concern is that you gave it ten out of possible ten points in the "Technical Feasibility" category…
> PC: Yes, and what is the problem?
> M: Well, this kind of score in this category implies that it would be a fairly simple project that can be done by our internal resources only.
> PC: I think we can handle it internally…
> M: We had several other new system implementations projects recently. Remember, the ones for the risk management team and for the HR department. In both cases we had to rely on extensive support from the external consultants provided by the vendors…I am not insisting on anything, but you will have to go in front of the executive committee and justify these projections.
> PC: You are right, let us downgrade the score from ten out of ten to five out of ten.

Summary

We started this chapter with the analysis of various types of portfolio appraisals: project selection, phase-level, and periodic project status reviews. As a part of this section, we also discussed the importance of considering organizational internal resource costs and a quick method to assess the throughput capacity of the organization's project pipeline.

Later, we examined the importance of mission and strategy and their impact on the portfolio management. Furthermore, this chapter discussed the importance of executive support and active involvement in all the aspects of portfolio management.

Also, we discussed the purpose and the recommended contents of the project and portfolio management charter that is used to outline all of the key project management and PPM processes of the organization.

Furthermore, this chapter described the "halo effect" that may occur during the creation of the scoring models and how to avoid it. We also examined the template and an actual sample of the business case document used to assess the project proposal.

Finally, we outlined the importance of portfolio monitoring and the significant roles project managers and PMO play in this process.

References

Appleseed Partners and OpenSky Research. Fourth product portfolio management benchmark study: The state of product development in 2013—A need for speed and a roadmap to sustainable innovation. Plainview, Austin, TX, 2013.

Boehm, B. *Software Engineering Economics.* Englewood Cliffs, NJ: Prentice-Hall, 1981.

Charan, R, Colvin, G. Why CEOs fail. *Fortune,* 139(12) (June 21, 1999) 68–72, 74–76, 78.

CIA World Factbook. Central Intelligence Agency, June 23, 2014. Web. September 10, 2014.

Cooper, RG, Edgett, SJ, Kleinschmidt, EJ. New product portfolio management: Practices and performance. *Journal of Product Innovation Management,* 16(4) (1999) 333–351.

Cooper, RG, Edgett, SJ, Kleinschmidt, EJ. *Portfolio Management for New Products.* New York: Basic Books, 2002.

Cooper, RG, Edgett, SJ, Kleinschmidt, EJ. *Best Practices in Product Innovation: What Distinguishes Top Performers.* Ancaster, Ontario, Canada: Stage-Gate International, 2003.

Deloitte Financial Services Group. *Elements for Successful Growth in Financial Services: Poised for Opportunities,* London, U.K.: Deloitte Financial Services Group, 2013.

Ernst and Young. The shifting pharmaceutical industry landscape: Accounting and regulatory trends affecting reporting for 2012 and planning for 2013. 2012: http://www.ey.com/Publication/vwLUAssets/The_shifting_pharmaceutical_industry_landscape/$FILE/The_shifting_pharmaceutical_industry_landscape.pdf. Accessed June 15, 2013.

Festinger, L, Riecken, HW, Schachter, S. *When Prophecy Fails.* Minneapolis, MN: University of Minnesota Press, 1956.

Flyvbjerg, B. Delusions of success: Comment on Dan Lovallo and Daniel Kahneman. *Harvard Business Review,* 81(12) (December 2003) 121–122.

Gartner. Gartner says worldwide IT spending forecast to reach $3.7 trillion in 2013. January 3, 2013.

Insight Research Corporation. The 2013 telecommunications industry review: An anthology of market facts and forecasts. Market Research Report, February 4, 2015.

Joint Research Centre. *The 2011 EU Industrial R&D Investment Scoreboard.* Brussels, Belgium: European Commission, 2011, p. 32.

KPMG Government and Public Sector Services. *Cutting through Complexity for the Public Sector and Government.* Amstelveen, the Netherlands: KPMG, July 2011.

Lovallo D, Kahneman, D. Delusions of success. How optimism undermines executives' decisions. *Harvard Business Review,* 81(7) (July 2003) 56–63, 117.

Mankins, MC, Steele, R. Turning great strategy into great performance. *Harvard Business Review*, (July–August 2005) 64–72.

Moustafaev, D. *Project Scope Management: A Practical Guide to Requirements for Engineering, Product, Construction, IT and Enterprise Projects*. New York: Auerbach Publications, 2014.

Moustafaev, J. *Delivering Exceptional Project Results*, 1st edn. J. Ross Publishing, Plantation, FL, 2010.

PMI. *A Guide to the Project Management Body of Knowledge*, 5th ed. Newtown Square, PA: PMI, 2012.

Project Management Institute. *2012 PMI Pulse of the Profession: Driving Success in Challenging Time*. Newton Square, PA: Project Management Institute, 2012.

Qfinance. Professional Services. 2013.

Rifkin, J. *The Third Industrial Revolution*. New York: Palgrave MacMillan, 2011.

Yakimov, G, Woolsey, L. *Innovation and Product Development in the 21st Century*. Gaithersburg, MD: Hollings Manufacturing Extension Partnership Advisory Board, 2010.

Bibliography

Bible, M. *Mastering Project Portfolio Management: A Systems Approach to Achieving Strategic Objectives.* Plantation, FL: J. Ross Publishing, 2011.

Bonham, S. *IT Project Portfolio Management.* Norwood, MA: Artech House Publishers, 2004.

Cooper, C, Martin, BK. Airport without planes shows Japan hooked on 'Useless Projects'. Bloomberg: http://www.bloomberg.com/apps/news?pid=newsarchive&sid=aYomskNBwinE. Accessed December 3, 2008.

Cooper, RG. Winning at new products: Pathways to profitable innovation. *PMI Research Conference 2006 Proceedings.* Montreal, Quebec, Canada, July 16–19, 2006.

Cooper, RG, Edgett, SJ, Kleinschmidt, EJ. Best practices for managing R&D portfolios. *Research Technology Management,* 41(4) (1998) 20–33.

Cooper, RG, Edgett, SJ, Kleinschmidt, EJ. Portfolio management for new product development: Results of an industry practices study. *R&D Management,* 31(4) (October 2001) 361–380.

Dawson, R. *The Seven MegaTrends of Professional Services: The Forces That Are Transforming Professional Services Industries and How to Respond.* San Francisco, CA: Advanced Human Technologies, 2006.

Deloitte. Alternative thinking 2013. Renewable energy under the microscope. Deloitte, New York, 2012a.

Deloitte. Oil and gas reality check. A look at the top issues facing the oil and gas sector. Deloitte, New York, 2012b.

Deloitte. *Technology, Media and Telecommunications Predictions 2013.* New York: Deloitte, 2012c.

Deloitte. The state of the global mobile consumer. Connectivity is core. Deloitte, New York, 2013.

Eggers, WD, Jaffe, J. *Gov on the Go: Boosting Public Sector Productivity by Going Mobile.* New York: Deloitte, 2013.

International Energy Agency. *World Energy Outlook 2012.* Paris, France: International Energy Agency, 2012.

Letavec, C. *The Program Management Office: Establishing, Managing and Growing the Value of a PMO.* Plantation, FL: J. Ross Publishing, 2006.

Levine, H. *Project Portfolio Management: A Practical Guide to Selecting Projects, Managing Portfolios and Maximizing Benefits*. New York: Jossey-Bass, 2005.

Moore, S. *Strategic Project Portfolio Management: Enabling a Productive Organization*. New York: Wiley, 2009.

Rothman, J. *Manage Your Project Portfolio: Increase Your Capacity and Finish More Projects (Pragmatic Programmers)*. Frisco, TX: Pragmatic Bookshelf, 2009.

Index